VHDL: HARDWARE DESCRIPTION AND DESIGN

VHDL: HARDWARE DESCRIPTION AND DESIGN

by

ROGER LIPSETT

CARL F. SCHAEFER

CARY USSERY

Intermetrics, Inc.

KLUWER ACADEMIC PUBLISHERS
Boston/Dordrecht/London

Distributors for North America:
Kluwer Academic Publishers
101 Philip Drive
Assinippi Park
Norwell, Massachusetts 02061 USA

Distributors for all other countries:
Kluwer Academic Publishers Group
Distribution Centre
Post Office Box 322
3300 AH Dordrecht, THE NETHERLANDS

Consulting Editor: Jonathan Allen, Massachusetts Institute of Technology

Library of Congress Cataloging-in-Publication Data

Lipsett, Roger, 1950–
    VHDL: Hardware description and design.

    Bibliography: p.
    Includes index.
    1. Computer input-output equipment—Computer
simulation.   2. VHDL (Computer program language)
I. Schaefer, Carl F., 1945–   . II. Ussery, Cary,
1962–   . III. Title.
TK7887.5.L57   1989   621.39 '2   89-11114
ISBN 0-7923-9030-X

Table of Contents

Chapter 7 - Large Scale Design 145

Chapter 8 - A Complete Example 177

Foreword

VHDL is a comprehensive language that allows a user to deal with design complexity. Design, and the data representing a design, are complex by the very nature of a modern digital system constructed from VLSI chips. VHDL is the first language to allow one to capture all the nuances of that complexity, and to effectively manage the data and the design process. As this book shows, VHDL is not by its nature a complex language.

In 1980, the U.S. Government launched a very aggressive effort to advance the state-of-the-art in silicon technology. The objective was to significantly enhance operating performance and circuit density for Very Large Scale Integration (VLSI) silicon chips. The U.S. Government realized that in order for contractors to be able to work together to develop VLSI products, to document the resulting designs, to be able to reuse the designs in future products, and to efficiently upgrade existing designs, they needed a common communication medium for the design data. They wanted the design descriptions to be computer readable and executable. They also recognized that with the high densities envisioned for the U.S. Government's Very High Speed Integrated Circuit (VHSIC) chips and the large systems required in future procurements, a means of streamlining the design process and managing the large volumes of design data was required. Thus was born the concept of a standard hardware design and description language to solve all of these problems.

In 1983, the U.S. Government issued a Request for Proposal to develop the language and to implement a set of tools for use with the language. The winner of the competitive bidding was a team composed of Intermetrics, International Business Machines, and Texas Instruments. As the lead engineer for the IBM part of the development effort, and as the chairman of the IEEE Computer Society Design Automation Standards Subcommittee, I saw directly the benefits that would accrue to industry from a standard hardware description language. VHDL is a descriptive language, one that is human and machine readable, so that it is well suited to preserve design specifications. VHDL is also a leading edge design language, one that may be used with many types of design tools such as simulators, design synthesizers, silicon compilers, placement and wiring tools, test generators, architectural specification and analysis tools, and timing analyzers, and can support research in all of these areas. At this time, many such tools are coming to the market place and university research in the named areas, which has been underway for years before the advent of VHDL, is proceeding to address itself to the use of VHDL. Why is this so?

As previously mentioned, a major power of VHDL is that it is a standard. Thus, industry can more easily communicate designs among participants in a design process. This ability to communicate designs is equally important in the research field, since, with VHDL, collaboration between researchers at various institutions becomes easier. But, standardization is not sufficient for the ground-swell of interest in the language. VHDL also has to be technically excellent and capable of allowing designers and researchers to describe the concepts they are developing and utilize the descriptions with tools in a way that simplifies the design or research process. VHDL achieves these goals.

The scope of VHDL covers the description of architectural description to gate level description. The language is hierarchical and mixed-level simulation is supported. The concepts embodied in the timing model for the language mirror real hardware -- the VHDL models of designs behave like real hardware. Many other excellent languages cover subsets of the capabilities that exist within VHDL, but none are as comprehensive. They do not cover the wide range covered by VHDL. Equally important, because VHDL is an IEEE standard, the language will have a significant effect on life-cycle support of products described in VHDL. At high levels of abstraction, the language makes an excellent specification medium for future designs to be created in new technologies or with alternative architectures. At lower levels of abstraction, the language serves well as a specification of what is to be fabricated.

All in all, VHDL is a "language for all seasons," one that supports design automation research, design and test, and product life-cycle. I take my hat off to the scores of people that have participated in the development of the VHDL concept and had the foresight to help bring it to fruition. I also salute the authors of this book for helping to bring understanding in the use of VHDL.

Ronald Waxman
University of Virginia
Charlottesville, Virginia
March 20, 1989

Preface

The VHSIC Hardware Description Language (VHDL) is a new hardware description language developed, starting in 1981, by the Very High Speed Integrated Circuits (VHSIC) Program Office of the Department of Defense for use as a standard language in the microelectronics community. This language represents a new step in the evolution of language support for hardware design. The recognized need for managing the complexity of information needed for digital design has driven the development of VHDL.

The acceptance process for a new language can be slow, especially for a language as rich in features as VHDL. The largest factor in this process is the dissemination of information about what the language is and, more importantly, how to use it. To date, little information has been available to the design community to explain what VHDL is, how it relates to what designers know, and how it can be integrated into the design process.

The authors of this book have, since the beginning of the VHDL program, been involved in developing the language, building tools that support the language, and using the language to build hardware descriptions. Intermetrics, Inc., for whom the authors work, was the Government's prime contractor for the original development of VHDL. Both Intermetrics and the authors are committed to the widespread use of VHDL as a hardware description language in the design community. This book represents an opportunity to inform; to share the knowledge of VHDL we have gained over the past five years.

This book describes the VHDL language and discusses ways in which VHDL can be used. It is intended to introduce the reader to all aspects of VHDL and to specific language constructs. It is not theoretical in nature and does not rigorously discuss the formal definition of the language; this approach is left to the Reference Manual for the language. What the book does do is take a look at the language feature by feature to provide an understanding of how each feature works and how it is integrated into the modeling process. This book is intended to give hardware designers and software model developers a thorough understanding of the VHDL language both as a language and as a design tool. We do not suggest particular ways of designing given hardware devices, but rather seek to provide the designer with an understanding of VHDL sufficient to allow him or her to best choose how to perform design.

The book is suitable for working professionals as well as graduate or advanced undergraduate study. Designers can view this book as a way to get acquainted with VHDL and as a reference for work with VHDL, while students and professors will find the book useful as a teaching tool for hardware design and hardware description languages in general, as well as VHDL in particular.

The presented material assumes a working knowledge of digital hardware design, as well as some familiarity with a high level programming language such as C, Pascal, or Ada.

Overview of the Book

The book is divided into ten chapters.

The first chapter is an introductory chapter which gives a brief history of VHDL, examines reasons for using VHDL and discusses terminology and conventions used throughout the book. The second chapter introduces the model of digital devices on which VHDL is based. It discusses the main semantic content of the language without going into the syntax of the constructs used within the model.

Chapter 3 introduces some basic elements of VHDL. These elements will be used throughout the rest of the book. Chapter 4 examines data types in the language in detail. Some readers might want to skim this chapter at first and then return to it as the sophistication of their VHDL models evolves to require the inclusion of user-defined data types.

The next three chapters, Chapters 5, 6, and 7, discuss VHDL from three viewpoints. Chapter 5 discusses using VHDL to describe the behavior of a hardware device. Chapter 6 discusses using VHDL to describe the structure of a hardware device. Chapter 7 discusses combining these two viewpoints into a model of a large hardware device. All these views of modeling are supported by the language. These chapters introduce the reader to the bulk of the VHDL language from a functional perspective instead of from a language definition perspective.

Chapter 8 ties the concepts presented in the previous chapters into a single design example. The design is a traffic light controller which is developed from a high level specification into a PLA implementation.

Chapter 9 looks at some of the more advanced features of the language. These features are used for high level modeling, test vector input, and other specialized functions.

Chapter 10 works through a few design examples as a way of illustrating the use of VHDL in system modeling and abstract design. Other books which treat VHDL tend toward the detailed gate level or board level modeling techniques. Here we try to illustrate some of the higher level features of VHDL and its uses in abstract modeling.

Acknowledgements

Many people gave us support, advice, and commentary on the earlier drafts of this book. In particular, our peers at Intermetrics provided extensive assistance and criticism. Special thanks are due to Al Gilman, Doug Dunlop, Victor Berman, and, especially, Kathy McKinley. They were all subjected to questionable prose and countless technical errors which they handled gracefully, and with beneficial results.

We would like to express our appreciation for the support and resources provided by Intermetrics, Inc. This book was developed on three different computers in three different cities and would not have been possible without the cooperation of Intermetrics. We would especially like to thank Bill Carlson, Victor Berman and Bill Bail at Intermetrics for the direct support they have given this book.

We would like to thank Carl Harris at Kluwer Academic and Rachael Rusting at Intermetrics, whose patience and forbearance have helped to make this book a reality and to motivate the authors. Paula Gillis of Intermetrics is responsible for the cover art. Peter Connor contributed time and ideas during the early stages.

A number of people have provided valuable comments on drafts of this book. Dr. David Hemmendinger and John van Tassel of Wright State University, Capt. John Evans of the Air Force Foreign Technology Division, and Lt. Karen Serafino of the Electronic Technology Laboratory at the Wright Research and Development Center deserve special recognition for their careful review of the text and the analysis and simulation of all the examples.

Over the past five years, many people have been involved with VHDL in one form or another. While we cannot mention all of these people here, we salute the effort and creativity of those who helped to bring VHDL to life. A few individuals deserve attention for keeping VHDL alive and well over the course of the past 5 years: Allen Dewey, who was the original Project Officer for the Air Force at Wright Field; Dr. John Hines, who has been a key factor in getting VHDL going and keeping the air in its sails; and Larry Saunders who, in addition to being a major proponent of VHDL in and out of IBM, chaired the IEEE VHDL Analysis and Standardization Group which carved out the current VHDL standard.

VHDL: HARDWARE DESCRIPTION AND DESIGN

Chapter 1
Introduction

The principal objective of the US Government's VHSIC program was to advance the state of the art in the United States in the areas of design, process, and manufacturing technology. The Government's objectives in sponsoring the VHDL effort as part of that program were to ensure that the VHSIC program would also provide the means to insert these new, advanced, microelectronic components into operational systems more rapidly than had been done in the past, and would provide technology that would allow more cost-effective development of commercial electronics by improving communication within and between companies, and by streamlining the development process. The VHDL effort formally began in June 1981 with a workshop, held in Woods Hole, Massachusetts, to discuss how best to achieve these goals. The workshop, which was sponsored by the VHSIC program, included members of industry, government, and academia.

The result of this workshop was a clearly stated goal of developing a new, industry standard, hardware description language that was technology independent and met the various needs of the members of the microelectronics community. The Woods Hole workshop also produced a detailed set of requirements that formed the technical basis of the Government's formal Request for Proposals.

Following this workshop, the Air Force issued a draft request for proposal, which was reviewed by the microelectronics community. This request for proposal was amended and finalized in early 1983. In July of 1983, the development team of Intermetrics, IBM and Texas Instruments was awarded the contract to develop the new language and to implement the support software needed to use the language.

Throughout the initial development stage the language was submitted to extensive public review. Following these interim reviews, the final version of the language to be developed under the original contract, known as VHDL Version 7.2, was made available in August of 1985. In July 1986, Intermetrics delivered the initial suite of support software; in February of 1987, a completely revised tool set for version 7.2 was made available.

In March of 1986, the IEEE took on the effort of standardizing VHDL. The VHDL Analysis and Standardization Group was set up to review the language, with a goal of repairing known problems with the language and modifying the language where a broad consensus formed around the modifications. The Air Force fully supported this work by awarding a contract to Intermetrics to develop the support software for the new language standard, now known as IEEE-1076. IEEE-1076 was approved in December 1987. This standard is the current definition of VHDL and is the language addressed in this book.

Why VHDL

VHDL was developed to address a number of recurrent problems in the development, exchange and documentation of digital hardware. For instance, a typical delivery of hardware to the government would include tens of thousands of pages of documentation that needed to be sifted through during acceptance and testing and referred to throughout the maintenance life of the component. When the component needed to be replaced it took a large amount of effort to reconstruct the intended behavior of the component. A good hardware description language solves this problem because the documentation is executable and all elements are tied into a single model.

While there are many hardware description languages, there was, prior to VHDL, no accepted standard in the industry. Many of the existing languages are developed to serve the simulators that run them, and are often proprietary developments of particular companies. Others target a particular technology, design level, or design methodology. VHDL is technology independent, is not tied to a particular simulator or value set, and does not enforce a design methodology on a designer. What it does do is allow the designer the freedom to choose technologies and methodologies while remaining within a single language. No one can foresee the changes that will take place in digital hardware technology in the future. Therefore, VHDL provides abstraction capabilities that facilitate the insertion of new technologies into existing designs.

VHDL thus offers a number of benefits over other hardware description languages.

Public Availability: VHDL was developed under a Government contract and is now an IEEE standard. The Government has a strong interest in maintaining VHDL as a public standard. The advantages of this status are apparent; without it, many of the other benefits described below would not exist.

Design Methodology and Design Technology Support: VHDL was designed to support many different design methodologies (*e.g.*, top-down versus library-based) and design technologies (*e.g.*, synchronous versus asynchronous, or PLA versus random logic). In this way, the language manages to provide design assistance to organizations that operate in very different ways, and that have very different design needs. VHDL is appropriate for use by a CAD/CAE house that sells library-based design tools, and is equally appropriate for an aerospace company that does a large amount of ASIC design.

Technology and Process Independence: VHDL was designed to be independent of both technology and process. By this we mean that VHDL does not have embedded within it an understanding of particular technologies or processes. However, such information can be written using VHDL. Thus, a working (*i.e.*, simulatable) description of a system may be developed above the gate level, and then decomposed into gate-level implementations depending on the chosen implementation technology (*e.g.*, CMOS, nMOS, GaAs). The advantages of this capability for second sourcing are apparent.

Wide Range of Descriptive Capability: VHDL supports behavioral description of hardware from the digital system (*e.g.*, box) level to the gate level. One of the primary advantages of VHDL lies in the ability to capture the operation of a digital system on a number of these descriptive levels at once, using a coherent syntax and semantics across these levels, and to simulate that system using any mixture of those levels of description. It is therefore possible to simulate designs that mix high-level behavioral descriptions of some subsystems with detailed implementations of other subsystems in the model.

This facilitates the development of a design that correctly reflects the intention of the original specification of the system. During maintenance of the system, redesigns or alterations to the system can be verified by replacing the VHDL in the description and resimulating with the test set.

Many existing hardware description languages operate best at the logic and gate level; for this reason, hardware designers may be most familiar with the use of HDL's to support low-level logic design. While VHDL is perfectly suited to this level of description, it extends well beyond that level. Therefore, the larger examples in this book have been purposely chosen to show VHDL in contexts beyond the logic level of digital hardware.

Design Exchange: VHDL is a standard and, as a result, VHDL models are guaranteed to run on any system that conforms to that standard. This means that models developed at one location will run at other locations (and produce the same simulation results) whether or not both locations use the same VHDL tool sets. This has many implications for the use of VHDL. As VHDL models of common components become available, designers will simulate their designs with those model descriptions. In this way designers can try a number of different components and choose those that best fit the constraints and requirements of the system being designed. Another aspect of this capability is that teams can exchange high-level descriptions of the subsystems of a digital system, which then allows each subsystem to be developed independent of progress on the other subsystems.

For example, suppose an organization was developing a digital system and wanted to contract out a subsystem to another organization. The contracting organization could develop a high-level VHDL description of the system with the subsystems broken out into design entities. The high-level description would then be given to the contractor for development. The contractor could develop the subsystem in VHDL while all along having a simulatable description of the rest of the system. The contracting organization would also be able to develop its portion of the system and test without waiting for the subsystem to be developed. This greatly reduces the integration effort and reduces the time of the entire development significantly.

Large-Scale Design and Design Re-use: VHDL was modeled on a philosophy similar to that of many modern programming languages — that design decomposition aids are just as important as detailed descriptive capabilities when it comes to supporting the development of large designs being executed by multiperson teams. There are several elements of the language specifically targeted at this goal. Packages, configuration declarations, and the concept of multiple bodies exhibiting different implementations of an entity are all present in the language to support design sharing, experimentation, and design management.

These elements of the language also tend to support design re-use, since they promote the encapsulation of design information and the ability to be flexible in conforming to others' design conventions when using an already-designed part.

Government Support: The VHSIC Program Office developed VHDL with a clear vision of how it itself wanted to use VHDL. VHDL can be used throughout the life cycle of a digital system, and is likely to become a major component of the procurement process. The Department of Defense is currently requiring VHDL descriptions on all contracts that develop ASIC's. As more VHDL toolsets become available and these toolsets mature the government can be expected to extend its mandates into other kinds of hardware development contracts and to increase the scope of the VHDL requirement to include certain information in the description and to provide simulatable descriptions.

Terminology and Conventions

VHDL is a hardware description language and, therefore, VHDL descriptions are generally used to model hardware components and system; *i.e.* gates, chips, boards, etc. However, VHDL provides an abstract framework for describing hardware which is easily extended into other domains. Therefore, the term *digital device* will be used when referring to a system that is being described in VHDL. A digital device can range from a gate to a microprocessor to a complete system and beyond. The guiding factor is that the underlying system be based on a general stimulus-response model that uses discrete (*i.e.* non-continuous) values.

Throughout this book the person who uses VHDL will be referred to as the *designer*. This is not because VHDL is primarily for those who design discrete systems, but rather because the use of VHDL requires one to model the processes of a system and define the nature of the stimulus that is applied to that system. In this sense, the *designer* may be designing a hardware component, the tests of that hardware component, or the constraints to be placed on the model.

In the examples, VHDL reserved words, such as **entity** and **signal**, are given in bold. References to these reserved words in the text are bolded when the reference is specifically to the term itself, and not a general reference to the concept.

Chapter 2
A Model of Hardware

VHDL is a language for describing general digital hardware devices. It makes no assumptions about the technology of a device or the methodology used to design a device. Instead, an abstraction of the nature of digital hardware devices is used as the basis for the language. This abstraction includes behavioral, timing, and structural characteristics of digital devices. These characteristics are integrated into a single cohesive language which provides a broad range of options for the description and design of digital devices. It is the ability to create a uniform description of a digital device utilizing multiple levels of abstraction which gives VHDL its power and applicability.

This chapter will present a model of this abstraction of digital devices. This model is appropriate for descriptions of all devices from high-level, abstract models down to gate-level descriptions. While theoretical in nature, this model should provide a robust foundation for understanding the concepts presented throughout the rest of the book. After this discussion, it will be easier to understand the nature of VHDL and the VHDL constructs which are used to model discrete systems.

The general model on which VHDL is based consists of three interdependent models: a behavioral model, a timing model, and a structural model. These three models are integrated into a single language.

A Model of Behavior

Behavior can be defined as the functional interpretation of a particular system. This book will explore structural and behavioral aspects of modeling. This does not imply the existence of purely structural models or purely behavioral models. All models have a structure and a behavior. In some design automation systems and languages, the behavior is not accessible to the designer except through the use of predefined library items. In VHDL, however, behavior is incorporated directly into the language and the designer has the option of mixing structure and behavior anywhere inside the model.

A digital device is a discrete system; *i.e.* one which transforms discrete-valued input into discrete-valued output. It does this by performing a number of operations or transformations on the input values. The results of these operations are passed into other operations and, finally, into the output values. A graphical representation of this concept is shown in Figure 2.1.

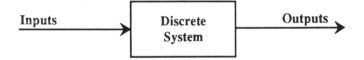

Figure 2.1. Representation of a Discrete System

As an example, consider a discrete system P whose behavior is the boolean function XOR. This system has two inputs, In1 and In2, and a single output, Out1. The discrete system is shown in Figure 2.2.

Discrete systems can range from the very simple to the very complex. For example, at the gate level operations are usually boolean functions applied to bit-valued wires and the discrete system is a representation of a logic circuit. On an abstract level, operations can be of any describable complexity. For instance, a microprocessor can be represented as a discrete system and the fetch cycle of that microprocessor could be a single operation.

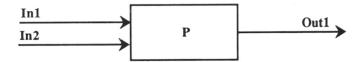

Figure 2.2. Representation of Discrete System P

In VHDL, all operations of a discrete system are described with the same abstract mechanism. Abstraction does not, however, imply complexity. The VHDL modeling mechanisms are essentially straightforward. As stated above, a discrete system is a collection of operations applied to values being passed through the system. In VHDL, we refer to each operation as a *process* and the pathways in which values are passed through the system as *signals*.

A process can be viewed as a program; it is constructed out of procedural statements and can call subprograms much as a program written in a general purpose procedural language like Pascal or C. In VHDL, all processes in a model are said to be executing concurrently. Therefore, a VHDL model is a collection of independent programs running in parallel.

In order to coordinate a set of processes running concurrently, a mechanism is defined to handle the communication between processes. This mechanism is the signal. In particular, signals define a *data pathway* between two processes. This data pathway is *directed*; one side of the pathway generates a value and one side receives that value. Data pathways have some special characteristics. First of all, each pathway has a type associated with it. The type defines a range of values which may be passed over the pathway. For instance, a data pathway whose type is integer could not contain a value of type bit, or a real number, or any value other than a value of type integer. All communication between processes takes place over these data pathways.

Processes continue to execute until they are suspended. Once suspended, a process can be reactivated. One way a process can be reactivated is by designating a maximum time for the process to remain suspended. When modeling behavior, a designer wants certain actions to take place when given conditions are met or necessary information becomes available. In particular, the designer often wants a process to be reactivated only when some change in the state of the system takes place. Such a change is reflected by a change in the value of a signal since signals contain the state of the system. VHDL provides a means for a process to express its sensitivity to the value of a data pathway. These pathways are called *sensitivity* channels. When the value on a sensitivity channel changes, the process is reactivated. The *wait statement* is used to designate any timeout conditions and sensitivity channels for a process. When a process executes a wait statement, the system records the reactivation conditions and the process is suspended.

As an example of reactivation conditions, consider the simple discrete system P described above. Figure 2.3 shows a representation of this discrete system which illustrates the data pathways and the internal process which implements the system. The discrete system should react whenever its inputs change, therefore, both inputs are also sensitivity channels for the process. This discrete system has a single operation.

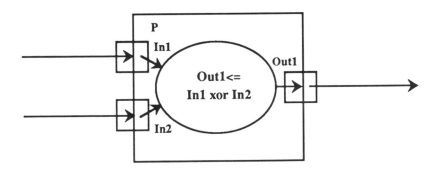

Figure 2.3. Discrete System P with Data Pathways and Process

The discrete system used to model a digital device is contained inside a *design entity*. A design entity has two parts: an entity declaration and an architectural body. The interface defines the inputs and outputs of the device and the body defines the discrete system used to model the device.

In many discrete systems, it is possible to decompose the behavior into a number of operations. For instance, the device depicted by discrete system P above could be represented by an equivalent discrete system Q which has three operations. The new discrete system is shown in Figure 2.4.

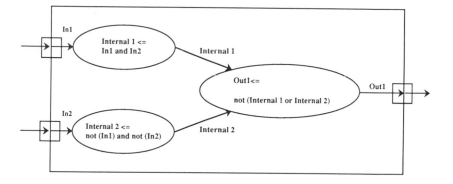

Figure 2.4. Representation of Discrete System Q

The discrete system Q has three operations which are functionally equivalent to the single operation of discrete system P above.

A Model of Time

The model of digital devices used here is based on a stimulus-response paradigm; when there is stimulus to the model, the model responds and then waits for more stimulus. This stimulus occurs at given times which are designated by the model of the discrete system. The "time" at which something occurs means simulation time and not the time of an implementation platform's internal clock. Since VHDL is concurrent but is also designed to be able to run on non-parallel machines it is necessary to create a definition of simulation time which defines when events occur during the course of simulation. Without such a definition, one model might simulate differently on two different VHDL simulators.

When a process generates a value on a data pathway it may also designate the amount of time before the value is sent over the pathway; this is referred to as *scheduling a transaction* after the given time. It is possible to schedule any number of transactions for a data pathway. The collection of transactions for a signal is called the *driver* of the signal. The driver is a set of time/value pairs which hold the value of each transaction and the time at which the transaction should occur.

VHDL has a two-stage model of time. This two-stage model is referred to as the *simulation cycle*. The simulation cycle is the conceptual abstraction under which a hardware model described in VHDL is exercised. This abstraction is based on a generalized model of the stimulus and response behavior of digital hardware; *i.e.* a functional component *reacts* to activity on its input connections and *responds* through its output connections. In VHDL, this model revolves around the concepts of processes and signals. Processes are the functional components which are connected with signals. A process may react to a change in value on a signal to which it is connected by transmitting new data to other processes via signals.

During the first stage of the simulation cycle, values are propagated through the data pathways (signals). This stage is complete when all data pathways which are scheduled to obtain new values at the current simulation time are updated. During the second stage, those active elements (processes) which receive information on their sensitivity channels are exercised until they suspend (via the execution of a wait statement). This stage is completed when all active processes are suspended. At the completion of the simulation cycle, the simulation clock is set to the next simulation time at which a transaction is to occur and the cycle is started again. The simulation cycle is depicted in Figure 2.5.

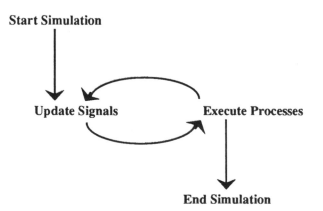

Figure 2.5. The Simulation Cycle

The above model means that there is always some delay between the time a process puts a value on a data pathway and the time at which the data pathway reflects that value. In particular, if no delay is given in the assignment of a value to the data pathway a *delta* delay is used. This delay does not update the time of the simulation clock but does require the passing of a new simulation cycle. It is important to take note of this fact because the language mechanisms for assigning values into a data pathway look deceptively like variable assignments in this and other languages but the effect is quite different. When a value is assigned in the pathway it is not immediately available to processes which read the value from the pathway. There is a *delay* between assigning and updating a signal value.

If the new value being assigned to a signal (*i.e.* passed through a data pathway) is different from the previous value of the signal, an *event* is said to have occurred. When a process is sensitive to a signal, it is sensitive to events on that signal, not to general transactions. If the process needs to become active when there is a transaction (*i.e.* even if the new value of the signal is the same as the old value) the process should be made sensitive to the signal-valued attribute 'Transaction which is described in Chapter 9.

A Model of Structure

When a model of a digital device requires more than a few operations, the representation of the associated discrete system can become difficult to manage. Such models often contain conceptual partitions which can be used to decompose the model into functionally related sections. This decomposition is the called the structure of the model.

Many digital devices are designed by combining a number of subdevices together and tying (wiring) the subdevices together. Each subdevice is a discrete system unto itself. A digital device which uses these subsystems is a new discrete system incorporating each of the subsystems.

The outermost data pathways of a discrete system are defined by the interface to the digital device which is defined by the entity. When a discrete system ties together two subsystems, it is really connecting a data pathway from one subsystem to the data pathway of the other subsystem. In this way, the two subsystems can communicate with each other. These connections are called *ports* and they have some special characteristics. First, the definition of a port represents a declaration of a signal and, therefore, of a data pathway. Think of data pathways as a pipe (in the plumbing sense) between two processes. The port represents an open end of the pipe. These open ends can be welded together with signals in a device which uses the subsystem containing the port.

For example, recall the discrete system Q shown earlier in the chapter. This discrete system required three operations. These three operations might be broken into two functional sections: stage 1 and stage 2. The resulting discrete system is shown in Figure 2.6.

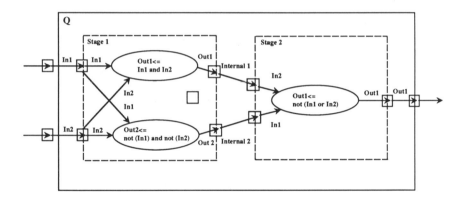

Figure 2.6. Decomposing Q into Two Subsystems

The functional sections are "black boxes" for the operations which they contain. By viewing these sections as black boxes it is possible to ignore the actual implementation of the operations.

The first box has two inputs and two outputs and contains the two processes which implement the first phase of the operation. The second box would have two inputs and a single output and would contain the process implementing the second phase of the operation.

The discrete system which wires these two subsystems together defines the data pathways (signals) used to connect the two subsystems and then ties the subsystems together. This has the effect of creating a single black box of the entire system. In a VHDL model of discrete system Q, each of the boxes would be represented by a VHDL design entity.

There are a number of ways in which the model of structure interacts with the model of behavior. First, it is possible to interject a type conversion function onto a data pathway. This is useful when two processes must communicate but the interfaces to the discrete systems in which they are defined do not have the same characteristics. For instance, one system might operate on integer signals and one system might operate on 32 bit encodings of integers. A type conversion could be used to convert between these two representations. Type conversion are discussed in detail in Chapter 7.

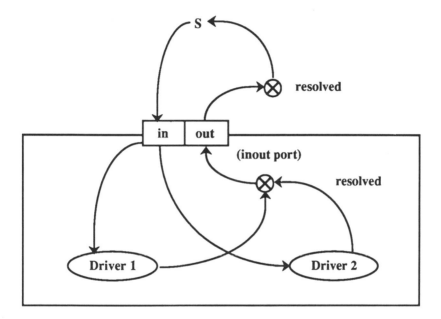

Figure 2.7. Resolution Over a Structural Hierarchy

Another important way in which structure and behavior interact is the resolution of resolved signals. If more than one driver is defined for a signal (*i.e.* more than one process assigns to the signal), the signal is a *resolved signal*. The designer must associate a resolution function with such a signal. The resolution function collects the values sent from each driver and generates a single value. For instance, a tri-state signal might be resolved with a function which performs a wired-or on the input values. In some situations, it is possible to *disconnect* a driver of a resolved signal; *i.e.* to *turn off* the driver. When a driver is turned off, it is not passed to the resolution function and, therefore, does not participate in the determination of the resolved value. There are two types of resolved signals: registers and busses. A register signal is *not* resolved if all of its drivers are turned off, while a bus signal is resolved when all of its drivers are turned off. This means that a resolution function for bus signals must handle the resolution of an empty list of values. If a port is resolved then the resolved value is determined before the value is passed out of the subsystem which contains the port. This is depicted in Figure 2.7.

Resolved signals have important implications for inout ports in a model. If a signal is resolved through a hierarchy of inout ports the value of the out side of each port is resolved all the way up to the top level signal and *then* passed down through each in side of the inout ports. This means that the value coming through the in side of the port is the value resolved at the top level signal and not the value resolved at each port. A detailed discussion of resolved signals can be found in Chapter 5.

Chapter 3
Basics

This chapter presents a broad view of some of the most important features of VHDL. The first section introduces the notions of structure and behavior and describes in very general terms how these notions relate to various VHDL features, including the VHDL signal, entity declaration, architecture body, process statement and component instantiation statement. VHDL permits designers to declare a virtually unlimited number of data types for characterizing the values held by signals, variables, and constants. The second section of this chapter explains how several basic types and objects are declared. VHDL's system of types is a complex subject; a detailed discussion of type and subtype declarations is contained in a separate chapter (Chapter 4). Modularity of design is a persistent theme in VHDL. The third section explains a fundamental mechanism underlying modularity: how a designer specifies interfaces to major constructs and how he connects constructs at their interfaces. The fourth section discusses the internal structure of several of the major VHDL constructs. The last section gives a brief overview of VHDL libraries and explains what determines a proper "order of analysis" for a set of library units.

Structure and Behavior

Two of the most basic questions we can ask about a microelectronic device are: How does it behave? and What does it consist of? The questions define two basic aspects of microelectronic hardware description: behavior and structure.

The basic concepts of hardware behavior are data transformation and timing. A data value at a point on a net may change from high to low (or from true to false, or from '1' to '0', or, at a more abstract level, from 10 to 20, etc.) in response to data transformations elsewhere in the hardware. There is always a delay of some magnitude (however small) between the causing transformation and the caused transformation.

The basic concepts of hardware structure are the *component*, the *port*, and the *signal*. The component is the building block of hardware description; virtually any "size" of building block — gate, chip, or board — can be seen as a component, depending on the level at which the hardware is being described. A port is the component's point of connection to the world, the point *through* which data flows into and out of the component.

The basic concept that ties the behavioral view and the structural view is the *signal*. As a descriptive abstraction of a hardware wire, the signal serves both to hold changing data values and to connect subcomponents. A signal is a path from one component to another component, the path *along* which data flows between components. Signals connect components at their ports.

In VHDL, a component is represented by the *design entity*. This is actually a composite consisting of an *entity declaration* and an *architecture body*. The entity declaration provides the "external" view of the component; it describes what can be seen from the outside, including the component's ports. The following entity declaration, which declares two **in** ports and two **out** ports, is the VHDL equivalent of the "external" view of a half-adder as shown in Figure 3.1:

```
-- The entity declaration
entity Half_adder is
  port (
    X : in Bit ;
    Y : in Bit ;
    Sum : out Bit ;
    Carry : out Bit) ;
  end Half_adder ;
```

("--" marks a comment in VHDL. Anything on a line following "--" is interpreted as comment.)

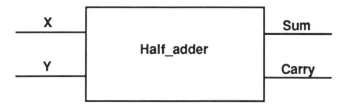

Figure 3.1. Half-adder: Entity Declaration

In an entity declaration, the entity itself and each of its ports is named with an *identifier*. A VHDL identifier is a sequence of characters (not containing any spaces) that obeys the following rules:

- Every character in the sequence is either a letter ('a' to 'z' and 'A' to 'Z'), or a digit ('0' to '9'), or the underscore character ('_').

- The first character in the sequence is a letter.

- The sequence contains no adjacent underscores, nor is the last character in the sequence an underscore.

VHDL identifiers are not case-sensitive; that is, an uppercase letter is considered to be the same letter as its corresponding lowercase letter. Thus, the following sequences of characters are considered to be variant forms of the same identifier:

        ABC
        abc
        aBc

Some identifiers, such as **begin, end,** and **entity**, are *reserved words* in VHDL. These identifiers, which are always printed in boldface in this book, have a fixed meaning in the language and may not be used to designate entities, signals, components, or other user-defined items.

The architecture body provides the "internal" view; it describes the behavior or the structure of the component. The following VHDL describes the behavior of the half-adder:

```
-- The architecture body:
architecture Behavioral_description of Half_adder is
begin
  process
    Sum <= X xor Y after 5 Ns ;
    Carry <= X and Y after 5 Ns ;
    wait on X, Y ;
  end process ;
end Behavioral_description ;
```

The architecture body has a *process statement* containing two *signal assignment statements* and a *wait statement*. The first signal assignment statement specifies that the signal connected to the port named Sum will obtain the "exclusive-or" of the value of the signal connected to the port named X and the value of the signal connected to the port named Y after a delay of 5 nanoseconds. Similarly, the second signal assignment statement specifies that the signal connected to the port named Carry will obtain the "and" of the values of the signals connected to the the ports named X and Y. (The left arrow, "<=", in the signal assignment statements is mnemonic for the direction of the data flow.) The wait statement suspends the process until there is an event on either of the signals X or Y, at which time execution resumes at the top of the process statement. The process statement thus describes the data transformation and the timing that together constitute the behavior of the half-adder.

Structure is described in VHDL by declaring signals and connecting the signals to the ports of subcomponents. The following example illustrates building a full-adder out of two half-adders and an OR gate (the corresponding drawings can be seen in Figures 3.2 and 3.3). The signal Temp_sum connects the sum output of the first half-adder to one input of the second half-adder. The signal Temp_carry_1 connects the carry output from the first half-adder to one input of the OR gate; the signal Temp_carry_2 connects the carry output from the second half-adder to the other input of the OR gate. These connections are specified in three *component instantiation statements*. Each of the three component instantiation statements names a component (Half_adder, Or_gate) that has been declared in a local *component declaration*.

```
entity Full_adder is
  port (
    A : in Bit ; B : in Bit ; Carry_in : in Bit ;
    AB : out Bit ; Carry_out : out Bit) ;
end Full_adder ;
```

architecture Structure **of** Full_adder **is**

-- signal declarations
signal Temp_sum : Bit ;
signal Temp_carry_1 : Bit ;
signal Temp_carry_2 : Bit ;

-- local component declarations
component Half_adder
  port (X : **in** Bit ; Y : **in** Bit ;
  Sum : **out** Bit ; Carry : **out** Bit) ;
end component;
component Or_gate
  port (In1 : Bit ; In2 : Bit ; Out1 : **out** Bit) ;
end component;

begin

-- component instantiation statements
U0: Half_adder
  port map (
    X => A, Y => B,
    Sum => Temp_sum, Carry => Temp_carry_1) ;

U1: Half_adder
  port map (
    X => Temp_sum, Y => Carry_in,
    Sum => AB, Carry => Temp_carry_2) ;

U2: Or_gate
  port map (
    In1 => Temp_carry_1, In2 => Temp_carry_2,
    Out1 => Carry_out) ;

end Structure ;

Each component instantiation statement is labeled with an identifier (U0, U1, U2) followed by a colon (:). Besides naming a component (Half_adder, Or_gate) declared in a local component declaration, a component instantiation statement contains an *association list* (this is the parenthesized list following the reserved words **port map**) that specifies which *actuals* (the signals Temp_sum, Temp_carry_1, and Temp_carry_2, and the ports A, B, Carry_in, AB, and Carry_out) are associated with which *locals* (the ports of the local component declaration). An actual associated with a local port may itself be either a signal or a port. VHDL signals and ports, although they are declared differently, both have projected waveforms (values scheduled to become current at future times), and are both said to belong to the class of signal objects (other classes of objects are *variables* and *constants*). Each association element in the example consists of the name of a local port followed by a right arrow (=>) and the name of an actual (signal or port).

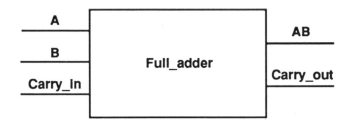

Figure 3.2. Full-adder: Entity Declaration

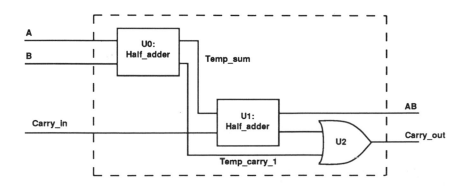

Figure 3.3. Full-adder: Architecture Body

Data Types and Objects

Something that can hold a value (a signal, for example) is an object. In VHDL, every object has a type, the type determining the kind of value the object can hold. In the preceding examples, all signals and ports have the type "Bit"; however, VHDL provides a designer with a rich variety of data types.

Data Types

If designers always described hardware at the lowest level of abstraction a single data type would suffice. One candidate for this all-purpose data type would be a three-level logic type with the values '0', '1', and 'Z'. However, there are many descriptive goals for which the three-level logic type would simply not be adequate. For example, a hardware description language should make it possible to specify the requirements for a floating point processor in terms of transformations on real numbers. Similarly, a hardware description language should make it possible to specify the top-level design of a microprocessor in terms more abstract than '0', '1' and 'Z'. VHDL was intended to support design at multiple levels and across multiple technologies; consequently, VHDL must provide a means for designers to define their own data types. But VHDL types are more than convenient tags for kinds of data values. The rules of VHDL stipulate that every object (signal, constant, or variable) and every expression has a single, uniquely determinable type and that (generally speaking) types cannot be mixed in expressions or in assignments of values to objects. For example, if a signal has been declared to have a certain floating point type (perhaps the signal represents a floating point register), then it is an error to assign the value of this signal to a second signal which has been declared to have a certain integer type (representing perhaps a general purpose register). The type of every object and expression can be determined *statically*; that is, the types can be determined prior to simulation — simply by analyzing the VHDL declarations and statements — and cannot change during the execution of the model. Thus types in VHDL provide a designer with a means of detecting design errors even before a design is simulated.

This chapter will cover only the most basic information on data types in VHDL; the following chapter will provide a more detailed treatment of the subject.

The three most basic VHDL data types are *integer types*, *floating point types*, and *enumeration types*. Integer types and floating point types are common to virtually all programming languages. The following examples illustrate how a designer would declare integer types:

         type Byte **is range** -128 **to** 127 ;
         type Bit_position **is range** 7 **downto** 0 ;
         type Decimal_int **is range** -1E9 **to** 1E9 ;

The *range constraint* in the declaration specifies the range of values that is defined for the type; values outside this range are not legal values for an object of this type. Floating point type declarations are similar in appearance but specify ranges with floating point numbers rather than with integers:

         type Fraction **is range** -1 + 0.1E-10 **to** 1 - 0.1E10 ;

Enumeration types have as their values *enumeration literals*. An enumeration literal is either an identifier or a *character literal*. Identifiers were discussed above in the section on entity declarations. A character literal is a single printable ASCII character enclosed in single quotes. The following are examples of character literals:

         'a'
         'A'
         ' '
         '''
         '%'
         '"'
         '6'

While identifiers are not case-sensitive, character literals are case-sensitive; thus the character literal 'a' is different from the character literal 'A'. An enumeration type declaration lists (between parentheses) the identifiers or character literals that are legal values of the type:

         type Three_level_logic **is** ('0', '1', 'Z') ;

VHDL has a small number of *predefined* types, that is, types which are available for use by the designer without the designer having to explicitly declare them. Among these predefined types are the integer type "Integer", the floating point type "Real", and the enumeration types "Boolean", "Bit", "Severity_level", and "Character". Here are the definitions of these types:

        -- VHDL allows each implementation to define the range of

```
-- Integer differently, but the range must be at least
-- -(2**31-1) to (2**31-1)
type Integer is range  -2147483648  to  2147483647 ;

-- VHDL allows each implementation to define the range of
-- Real differently, but the range must be at least
-- -1E38 to 1E38, and the implementation must support
-- at least six decimal digits of precision.
type Real is range
  -16#0.7FFF_FF8#E+32  to  16#0.7FFF_FF8#E+32 ;

-- Type Boolean is really an enumerated type, not a primitive
type Boolean is (False, True) ;

-- Both Boolean and Bit are two-valued enumeration types, but
-- their literals are distinct.
type Bit is ('0', '1') ;

-- Type Severity_level is used in assertion statements.
type Severity_level is (Note, Warning, Error, Failure) ;

-- Type Character has one literal for each of the
-- 128 ASCII characters; the nonprintable characters
-- are represented by identifiers.
type Character is (
    NUL, SOH, STX, ETX, EOT, ENQ, ACK, BEL,
    BS,  HT,  LF,  VT,  FF,  CR,  SO,  SI,
    DLE, DC1, DC2, DC3, DC4, NAK, SYN, ETB,
    CAN, EM,  SUB, ESC, FSP, GSP, RSP, USP,

    ' ', '!', '"', '#', '$', '%', '&', ''',
    '(', ')', '*', '+', ',', '-', '.', '/',
    '0', '1', '2', '3', '4', '5', '6', '7',
    '8', '9', ':', ';', '<', '=', '>', '?',

    '@', 'A', 'B', 'C', 'D', 'E', 'F', 'G',
    'H', 'I', 'J', 'K', 'L', 'M', 'N', 'O',
    'P', 'Q', 'R', 'S', 'T', 'U', 'V', 'W',
    'X', 'Y', 'Z', '[', '\', ']', '^', '_',

    '`', 'a', 'b', 'c', 'd', 'e', 'f', 'g',
    'h', 'i', 'j', 'k', 'l', 'm', 'n', 'o',
    'p', 'q', 'r', 's', 't', 'u', 'v', 'w',
    'x', 'y', 'z', '{', '|', '}', '~', DEL ) ;
```

In addition to these predefined types, the predefined type "Time" deserves mention among the basic VHDL types. The declaration of type "Time", which is discussed in Chapter 4, allows a user to specify time expressions in units such as picosecond, nanosecond, microsecond, etc.

Objects

There are three classes of objects in VHDL: signals, variables, and constants. A signal and a variable can both be assigned a succession of values, while a constant is assigned a value one time only (when it is declared). A variable differs from a signal in that the value assigned to a signal does not become the current value until some time in the future while the value assigned to a variable becomes the current value immediately. Unlike signals, which are analogous to wires in hardware, variables have no direct analogs in hardware. Rather, variables are useful in doing the computations required for higher-level modeling of hardware behavior. VHDL is a strongly typed language; every object has a type and can only hold values of that type. The type of an object is fixed in its declaration. The following are examples of object declarations:

> **constant** ROM_size : Integer := 16#FFFF# ;
> **variable** In_fetch : Boolean ;
> **signal** Enable : Bit ;
> **signal** CLK, CLEAR : Bit := '0' ;
>   -- declares two signals
> **variable** Address : Integer **range** 0 **to** ROM_size ;
>   -- declares a variable of type Integer with
>   --   an added range constraint

The general form of an object declaration is

> *object-class identifier-list* : *subtype-indication signal-kind*
>   := *expression*

The object class is either **constant, variable,** or **signal.** The identifier list is one identifier or multiple identifiers separated by commas (,). The subtype indication specifies the type of the objects declared by the object declaration; the subtype indication may simply name a type or, as in the case of the variable ROM_size, it may name a type with an additional constraint (constraints on types are discussed in Chapter 4). The signal kind is either **bus** or **register**; the signal kind is optional and may only appear in a signal declaration. The significance of **bus** and **register** will be explained in Chapter 5. The colon-equal (:=) and expression specify an initial default value; this is optional in an object declaration.

VHDL has a large number of rules relating to type compatibility. Among the more important of these rules are the following:

- The type of an expression assigned to an object must be the same as the type of the object.

- The operands of many predefined operators (e.g. arithmetic and relational operators) must be of the same type.

- The type of an actual must be the same as the type of the formal it is connected to.

The declarations of "ROM_size", "In_fetch", and "Enable" in the immediately preceding examples were *explicit object declarations*. But there are other kinds of objects in VHDL, including the following:

- Ports. All ports are of the object class signal. Ports can be declared in entity declarations, component declarations, and block statements. (Entity declarations are discussed below. Component declarations are discussed in Chapter 6. Block statements are discussed in Chapter 7.)

- Generics. Generics are a way of passing environment information to subcomponents; they can be declared in the same constructs in which ports can be declared. All generics are of the object class constant. (Generics are discussed in detail in Chapter 6.)

- Parameters. Parameters are declared on subprograms (functions and procedures). Parameters on functions are of the object class constant; parameters on procedures may be of any object class. (Subprograms are discussed in a later section of this chapter.)

- Indexes on loop statements and generate statements. (Loop statements are discussed in Chapter 5; generate statements are discussed in Chapter 6.)

Hooking Constructs Together

The example of the full-adder involved an *interface list* that declared local ports in a component declaration and an *association list* in a component instantiation statement that associated actuals with the local ports. Interface lists and association lists are very basic features of VHDL; essentially, they are the means that a designer uses to hook together separate constructs in the language.

Interface Lists

There are four VHDL constructs that specify *interfaces*, or potential points of communication between separate units. These are the entity declaration, the local component declaration, the block statement, and the subprogram (function or procedure) specification. An *interface list* is a parenthesized list of *interface elements*, separated by semicolons (;). Each interface element declares one or more interface objects, and each such object is a separate potential point of communication. The objects in a single interface element have three things in common:

- Object class (**signal, constant,** or **variable**)
- Mode (direction of data flow: **in, out, inout, buffer**)
- Data type (for example, "Bit"; data types are discussed briefly in this chapter and more fully in the next chapter)

Each entity declaration, each component declaration, and each block statement has (possibly) two interface lists: one for ports and one for generics. A subprogram specification has one interface list for subprogram parameters. The ports and generics of an entity declaration or of a block statement are referred to as *formal ports* and *formal generics*; the ports and generics of a local component declaration are referred to as *local ports* and *local generics*. The interface objects of a subprogram specification are referred to as *formal subprogram parameters* or simply *formal parameters*. The following table shows which modes and classes are allowed for each of the three types (port, generic, parameter) of interface object:

Interface Lists in Entities, Components, Blocks		
Interface Object	Class of Object	Allowed Modes
port	signal	**in out inout buffer**
generic	constant	**in**

Interface Lists in Subprograms		
Interface Object	Class of Object	Allowed Modes
parameter	signal	**in out inout**
parameter	constant	**in**
parameter	variable	**in out inout**

The general form for a single interface element in an interface list is:

object-class identifier-list : *mode subtype-indication* **bus**
   := *expression*

As in ordinary object declarations (see above), the *object class* is either
constant, signal, or **variable.** The *mode* is either **in, out, inout,** or
buffer (only a port can be of mode **buffer; buffer** ports will be
explained in Chapter 6). The object class and/or the mode of an
interface element may be omitted; the defaults that apply are discussed
under the various kinds of inteface lists. The reserved word **bus** is
optional but may appear only if the object class is **signal; bus** signals
are discussed in Chapter 5. The colon-equal (:=) and expression together
indicate a default initial value to be applied to the objects declared by
the interface element; they are optional. Here are some example
interface elements:

    -- the minimal amount of information
    X : Bit

    -- with more than one identifier
    X, Y : Bit

    -- with class, mode, and default initial value
    constant X, Y : **in** Bit := '0'

    -- with a subtype indication consisting of a type and a
    -- range constraint
    signal X, Y : **out** Integer **range** -100 **to** 100 **bus** := 0

As the last two examples show, several interface elements sharing the
same class, mode, type, and default initial value can be grouped into a
single interface element containing an *identifier list.* The identifiers in an
identifier list are separated by commas (,).

   The interface element is almost identical in form to the explicit
object declaration. There are two differences. First, a semicolon is
actually *a part of* each object declaration, while a semicolon appears
between two interface elements. Second, a signal declared in an object
declaration can have a signal kind of either **bus** or **register,** while an
interface signal (port or parameter) can only have the signal kind **bus.**

Association Lists

   There are four constructs that specify *associations,* or actual
communication paths between separate units. These are the component

instantiation statement, binding indication, block statement, and subprogram (function or procedure) call. (While a component instantiation statement specifies connections to a local component, a binding indication specifies connections to an entity; this will be explained in Chapter 6.) An *association list* is a parenthesized list of *association elements*, separated by commas (,). Each association element specifies a connection between an actual (an object or expression) in one unit and a formal or local in another unit. Each component instantiation statement, each binding indication, and each block statement has (possibly) two association lists: one for associating actuals with ports and one for associating actuals with generics. A subprogram call has one association list for associating actuals with parameters. Component instantiation statements are said to associate actuals with locals; subprogram calls, binding indications, and block statements are said to associate actuals with formals.

In the example of the full-adder, each association element was in a notation known as *named association*. In *named association*, the formal (or local) as well as the actual is specified: the formal (or local) occurs to the left of the right arrow (=>) and the actual occurs to the right of the right arrow. Since named association specifies unambiguously one formal (or local), it is not required that the order of associations be the same as the order in which the formals (or locals) were declared in the interface list. There is an alternate notation for an association element known as *positional association*. In *positional association*, an association element specifies only an actual; the formal (or local) is inferred by matching the ordinal position of the actual in the association list with the ordinal position of a formal (or local) in the interface list. It is possible for an association list to have some association elements in named notation and some in positional notation; however, no positional associations may follow a named association in an association list. Here is the first component instantiation statement from the full-adder example rewritten (with no change in meaning) in positional notation:

U0: Half_adder **port map** (A, B, Temp_sum, Temp_carry_1) ;

Major VHDL Constructs

This section discusses a number of the major building blocks available to the VHDL designer: the entity declaration, the architecture body, the subprogram, the package declaration, and the package body.

Entity Declarations

Examples of entity declarations were given at the beginning of this chapter. The general form of the entity declaration is

> **entity** *identifier* **is**
>   **generic** *interface-list* ;
>   **port** *interface-list* ;
>   *declarations*
> **begin**
>   *statements*
> **end** *identifier* ;

As mentioned earlier, generics are a way for an instantiating component to pass environment information to an instantiated subcomponent; formal generics will be discussed in Chapter 6. Formal ports were seen in the entity declarations of the half-adder and the full-adder and will be more fully discussed in Chapter 6.

The entity declaration is one of several major VHDL constructs that may themselves contain declarations. These major constructs are the entity declaration, architecture body, package declaration, package body, and subprogram body — all of which are discussed in this chapter — as well as the block statement and process statement, which are discussed in later chapters. There is a group of declarations, each of which can occur in each of these major constructs. This group, called the group of *basic declarations*, includes the following:

> *type declaration*
> *subtype declaration* (explained in Chapter 4)
> *constant declaration*
> *file declaration* (explained in Chapter 9)
> *alias declaration* (explained in Chapter 9)
> *subprogram declaration* (explained below in this chapter)
> *use clause* (explained below in this chapter)

Declarations that can follow the formal ports in an entity declaration include:

> *basic declaration*
> *signal declaration*
> *subprogram body* (explained below in this chapter)
> *attribute declaration* (explained in Chapter 9)
> *attribute specification* (explained in Chapter 9)

disconnection specification (explained in Chapter 9)

An entity declaration can also contain a set of *entity statements* after the reserved word **begin**. It may seem odd that an entity declaration, which is intended to represent only the interface to a component, should be allowed to contain statements. However, entity statements are not intended to represent a component's dynamic behavior; instead, entity statements are intended to represent checks or assertions about the component's interface. For example, an entity statement might be used to check that the values of a preset port and a clear port on a flip-flop are not both '0'. Placing this check in the entity declaration rather than in the architecture body (where the rules of VHDL would also allow it to be placed) means that the check will be present in all design entities built from the entity declaration and that the check need not be duplicated in each architecture body representing an alternative implementation of the component. Given this restricted function of entity statements, it is appropriate that the only kinds of statements that are allowed in entity declarations are:

> *concurrent assertion statement*
> *concurrent procedure call*
> *process statement*

(These statements are discussed in Chapter 5.) In keeping with the intention that entity statements do not represent dynamic component behavior, there is a further restriction on any concurrent procedure call or process statement occurring in an entity declaration: they must not assign values to any signal object (signal or port). A concurrent procedure call or process statement that does not assign to any signal object is called a *passive* concurrent procedure call or process statement.

Any of the following elements of the full form may be omitted in an entity declaration: the formal generic declarations; the formal port declarations; the other declarations; the reserved word **begin** and the concurrent statements; and the closing identifier. Thus, the following minimal entity declaration is legal VHDL:

> **entity** E **is**
> **end** ;

While this entity declaration might appear to describe a singularly uninteresting piece of hardware with no inputs and no outputs, it is not an entirely useless component. It could, for example, represent a self-contained testbench setup; the architecture of this empty entity might instantiate three subcomponents: a unit under test, a test vector generator (possibly one that read in values from a file using VHDL file I/O), and a comparator that checked actual results against predicted results.

Architecture Bodies

Examples of architecture bodies were given at the beginning of this chapter. The general form of the declaration of an architecture body is

> **architecture** *identifier* **of** *entity-mark* **is**
>   *declarations*
> **begin**
>   *statements*
> **end** *identifier* ;

Declarations that can occur in an architecture body include:

> *basic declaration*
> *signal declaration*
> *subprogram body*
> *attribute declaration*
> *attribute specification*
> *disconnection specification*
> *component declaration*
> *configuration specification* (discussed in Chapter 6)

Any concurrent statement may occur in an architecture body. Concurrent statements (which are discussed in Chapters 5 and 6) include:

> *concurrent signal assignment statement*
> *concurrent assertion statement*
> *concurrent procedure call*
> *process statement*
> *component instantiation statement*
> *block statement*
> *generate statement*

The declarations and/or statements may be omitted in an architecture body, but an architecture body with no statements has no apparent design function.

Subprograms

VHDL provides a number of predefined operators — such as "+" for addition, "<" for the less-than relation, "**xor**" for the logical exclusive-or operation — for the the construction of expressions that compute values. These and other predefined operators will be discussed in the following chapter. The predefined operators have a fixed number of operands whose types must conform to certain rules. But VHDL also allows a designer to define subprograms, which can be seen as a generalization of operators. Like the operators, they are defined for a fixed number of operands (parameters) whose types must conform to the types in the subprogram declaration. They are more general than operators in two respects. First, a subprogram may be devised for any number of operands (parameters) and any combination of types. Second, while there are subprograms that, like operators, merely compute values without affecting the values of objects or controlling the flow of execution, there are also subprograms (procedures) that assign new values to objects and even suspend execution.

A subprogram is a sequence of declarations and statements that can be invoked repeatedly from different locations in a VHDL description. VHDL provides two kinds of subprograms: procedures and functions. Procedures differ from functions in that the invocation of a procedure is a statement while the invocation of a function is an expression. VHDL distinguishes between a subprogram declaration and a corresponding subprogram body. The subprogram declaration contains only interface information, while the subprogram body contains interface information, local declarations, and statements. The distinction between subprogram declaration and subprogram body is analogous to the distinction between entity declaration and architecture body: it is a way of separating interface from algorithm (or, in the case of the architecture body, from structure).

A subprogram declaration is most conveniently described as consisting of a *subprogram specification* followed by a semicolon (;). There are two forms of subprogram specifications, one for a procedure and one for a function.

> **procedure** *identifier interface-list*
> **function** *identifier interface-list* **return** *type-mark*

Functions are intended to be used strictly for computing values and not for changing the value of any objects associated with the function's formal parameters; therefore, all parameters of functions must be of mode **in** and must be of class **signal** or **constant**. Procedures, on the other hand, are permitted to change the values of the objects associated with the procedure formal parameters; therefore, parameters of procedures may be of mode **in, out,** and **inout**. If no mode is specified for a subprogram parameter, the parameter is interpreted as having mode **in**. If no class is specified, parameters of mode **in** are interpreted as being of class **constant** and parameters of mode **out** and **inout** are interpreted as being of class **variable**. It is possible to define a procedure or function with no parameters; in this case, the interface list is simply omitted from the subprogram specification. Note that a function, unlike a procedure, must declare the type of the value the function will return. Examples of subprogram declarations follow:

```
-- procedure declaration with no parameters
procedure P ;

-- function declaration with no parameters
function Limit return Real ;

-- object class of parameter is implicit
procedure Mod_256 (X : inout Integer) ;
-- object class of parameter is explicit
procedure Mod_256 (variable X : inout Integer) ;

-- object class of parameter is implicit
function Mod_256 (X : Integer) return Byte ;
-- object class of parameter is explicit
function Mod_256 (constant X : in X) return Byte ;

-- parameters of object class signal
procedure Xor2(signal In1, In2 : Bit; signal Out1 : out Bit);
```

The full form of a subprogram (procedure or function) body is

```
subprogram-specification is
   declarations
begin
   statements
end identifier ;
```

The declarations and/or statements may be omitted. However, when a function is called, the function call must terminate by executing a *return statement*, which determines the value returned by the function call (return statements are discussed in Chapter 5). Hence, any function call to a function with an empty body will result in an execution error. The closing identifier (after **end**) is also optional. Any of the following declarations may occur in a subprogram body:

> *basic declaration*
> *variable declaration*
> *subprogram body*
> *attribute declaration*
> *attribute specification*

While a subprogram may call other subprograms, it may not instantiate components; therefore, a subprogram should not need to declare signals to wire together subcomponents. Note that a subprogram may itself declare "nested" subprograms.

Here is a simple subprogram body:

```
function Min (X, Y : Integer) return Integer is
begin
  if X < Y then
    return X ;
  else
    return Y ;
  end if ;
end Min ;
```

Packages and Use Clauses

Data types, constants and subprograms can be declared inside entity declarations and inside architecture bodies, and the types, constants, and subprogram declarations that are declared inside an entity declaration are visible in (can be used by) its associated architecture bodies. However, the data types, constants, and subprograms declared inside entity declarations and architecture bodies are not visible to other entity declarations or their architecture bodies. To make a set of type, constant and subprogram declarations visible to a number of design entities (entity declarations and their associated architecture bodies), VHDL provides the package. The general form of a package declaration is

> **package** *identifier* **is**
> *declarations*
> **end** *identifier* ;

Both the declarations and the closing identifier are optional; however, an empty package declaration does not serve any apparent design purpose.

The following declarations may occur in the declarations portion of a package declaration:

> *basic declaration*
> *signal declaration*
> *attribute declaration*
> *attribute specification*
> *disconnection specification*
> *component declaration*

The following example illustrates a package declaration.

package Logic **is**

  type Three_level_logic **is** ('0', '1', 'Z') ;
  constant Unknown_value : Three_level_logic := '0' ;
  function Invert (
    Input : Three_level_logic)
  return Three_level_logic ;

  end Logic ;

The general form of a package body is

package body *identifier* **is**
  declarations
end *identifier* ;

A package body always has the same name as its corresponding package declaration, preceded by the reserved words **package body**. Declarations that may occur in a package body include:

> *basic declaration*
> *subprogram body*

As with the package declaration, the package body may be empty, and its closing identifier is optional. Note that a package body contains the subprogram bodies whose corresponding subprogram declarations occur in the package declaration. The following is an example of a body for package Logic given above:

```
package body Logic is

  -- Subprogram body of function Invert
  function Invert (
    Input : Three_level_logic)
  return Three_level_logic is
  begin
    case Input is
      when '0' =>
        return '1' ;
      when '1' =>
        return '0' ;
      when 'Z' =>
        return 'Z' ;
    end case ;
  end Invert ;

end Logic ;
```

Package bodies contain subprogram bodies and declarations that are not intended to be used by other VHDL units. Thus, package declarations contain the public, visible declarations while package bodies contain the private, invisible declarations.

Items declared inside a package declaration are not automatically visible to another VHDL unit. A use clause preceding a unit will make items declared in a package declaration visible in the unit.

```
  -- The declaration of package Logic (above) is assumed
  -- The following use clause makes the declarations of
  -- Three_level_logic and Invert visible to the entity declaration
  use Logic.Three_level_logic, Logic.Invert ;
  entity Inverter is
    port (X : in Three_level_logic ; Y : out Three_level_logic) ;
  end Inverter ;

  -- The architecture body inherits the visibility of its entity
  -- declaration, so the use clause does not need to be repeated.
  architecture Inverter_body of Inverter is
  begin
    process
    begin
      Y <= Invert (X) after 10 ns; -- A function call
      wait on X ;
    end process ;
  end Inverter_body ;
```

A **use** clause may, as in the immediately preceding example, name specific items declared in a package. However, an alternate form of the **use** clause, constructed with the reserved word **all**, makes all items declared in a package declaration visible.

```
-- The following use clause makes all declarations
-- in package Logic visible to the entity declaration.
use Logic.all ;
entity Inverter is
  port (X : in Three_level_logic; Y : out Three_level_logic);
end Inverter;

-- The architecture body inherits the visibility of
-- its entity declaration
architecture Inverter_body of Inverter is
begin
  process
  begin
    Y <= Invert (X) after 10 ns ;
    wait on X ;
  end process ;
end Inverter_body ;
```

Additional discussion of packages and **use** clauses can be found in Chapter 7.

Libraries

VHDL allows a designer to maintain multiple design libraries, and the various pieces of a single design may reside in different libraries. This section explains the notion of a separately-analyzable library unit and explains how the names of library units and the names of libraries are made visible so that they can be referenced by other units. Additional discussion of libraries can be found in Chapter 7.

Library Units and Order of Analysis

Analysis is the process of checking a VHDL unit for syntactic and semantic correctness and inserting the VHDL unit (if it is correct) into a library. A library unit is a VHDL construction that can be separately analyzed. There are two classes of library units: primary units and secondary units. The primary units are entity declaration, package declaration and configuration declaration (configuration declarations will be explained in Chapter 7); the secondary units are package body and architecture body. In a given library, there can be at one time only one primary unit with a given name, but there may be multiple secondary units with the same name and there may be secondary units with the same name as a primary unit. Secondary units must be analyzed into the same library as their corresponding primary unit. A library unit is separately analyzable in the sense that the VHDL text for a library unit is not contained inside the text for another VHDL unit. However, "separately analyzable" does not mean "analyzable in any order". There are three rules governing the order in which library units may be analyzed. First, a secondary unit can be analyzed only after its corresponding primary unit has been analyzed. Second, any library unit that references a primary unit can only be analyzed only after that primary unit has been analyzed. Third, an architecture body referenced in a configuration declaration must be analyzed before the configuration declaration; likewise, one configuration declaration referenced in a second configuration declaration must be analyzed before that second configuration declaration.

Since the name of a package body is always the same as the name of its corresponding package declaration, there can never be more than one package body for a given package declaration. Architecture bodies are not subject to this same restriction. The name of an architecture body may be different from the name of its corresponding entity declaration, and it is possible to have multiple architecture bodies corresponding to a single entity declaration. This gives the designer the ability to maintain multiple hardware designs for a single hardware interface or to maintain abstract, behavioral implementations in addition to concrete, structural implementations. Figure 3.4 shows a sample library with five library units.

Visibility of a Primary Unit and Libraries

The name of a primary unit (for example, a package declaration) is not automatically visible to other units. A variant of the **use** clause (with the name of a library to the left of the dot and the name of a primary unit to the right of the dot) provides the needed visibility. The following example assumes that the package Logic has been analyzed into the library Basic_library.

```
-- The first use clause makes the name of package Logic
-- visible; the second use clause makes the declarations
-- in package Logic visible.
use Basic_library.Logic ;
use Logic.all ;
entity Inverter is
```

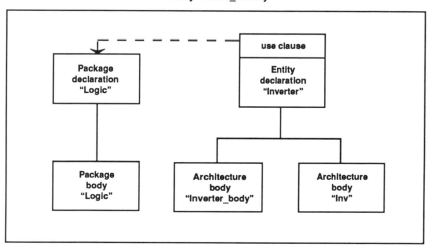

Library "Basic_Library"

Figure 3.4. Library Units

    port (X : **in** Three_level_logic ; Y : **out** Three_level_logic) ;
end Inverter ;

The alternate form of the use clause,

    use Basic_library.**all** ;

would make visible the names of all primary units in library Basic_library.

    Just as an item declared inside a package declaration is not automatically visible and just as the name of a primary unit is not automatically visible, so the name of a library is not automatically visible. But while a **use** clause provides the needed visibility in the cases of an item declared inside a package declaration and the name of a primary unit, it is a *library clause* that provides the needed visibility in the case of a library. The following VHDL and Figure 3.5 illustrate the use of multiple libraries.

-- The following library clause makes the name of library
-- Basic_library visible
library Basic_library ;
use Basic_library.Logic ;
use Logic.**all** ;
entity Inverter **is**
   port (X : **in** Three_level_logic ; Y : **out** Three_level_logic) ;
end Inverter ;

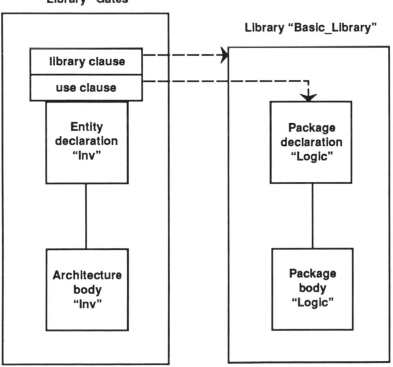

Figure 3.5. Multiple Libraries

    There is one exception to the rule that the name of library is not
automatically visible: the library name "Work" is always visible (just as
if the library clause "**library** Work;" actually preceded every unit).
Furthermore, the library name "Work" always designates the library
which the current library unit is being analyzed into. "Work" is thus a

logical name, whose interpretation is dependent on a particular installation and, within this installation, on a particular environment that exists at the time a library unit is analyzed.

A library name can be referenced in VHDL, but VHDL does not give the designer a means of creating a library or defining a library name. Since libraries must be defined outside of VHDL, the use of library names in a VHDL description would make the description more installation-dependent and less portable. The logical library name "Work" makes it possible to write VHDL descriptions that are portable across design environments.

Chapter 4
Data Types

VHDL permits designers to declare a virtually unlimited number of data types for characterizing the values held by signals, variables, and constants. The previous chapter mentioned briefly three data types: integer types, floating point types, and enumeration types. This chapter provides a detailed description of the other major kinds of data types in VHDL (two types not discussed in this chapter, access types and file types, are discussed in Chapter 9). This chapter consists of six sections. The first section of this chapter explains the various kinds of literals that are used in designating values of different types. The second section describes how scalar types are declared. The third section describes how composite types (arrays and records) are declared, how their values are expressed, and how their elements are referenced. The fourth section explains the notion of a subtype. The fifth section presents various predefined attributes that apply to types and array objects. The sixth section covers the various predefined operators that are used to construct expressions.

Literals

A literal is a symbol (other than a reserved word like **begin** or an operator symbol like "+") whose value is immediately evident from its representation. There are six kinds of literals: integer literals, floating point literals, character literals, string literals, bit string literals, and physical literals.

A floating point literal differs from an integer literal in that a floating point literal contains a dot (.) while an integer literal never contains a dot. Both floating point literals and integer literals may be expressed in any base from 2 to 16, and both kinds of literals may have exponents in base 10 (floating point literals, but not integer literals, may have negative exponents). Any two adjacent digits in an integer or floating point literal may be separated by a single underscore (_) character; this underscore is strictly to improve legibility.

2	-- base 10 integer literal
0	-- base 10 integer literal
77459102	-- base 10 integer literal
10E4	-- base 10 integer literal
16#D2#	-- base 16 (hex) integer literal
8#720#	-- base 8 (octal) integer literal
2#11010010#	-- base 2 (binary) integer literal
1.0	-- base 10 floating point literal
0.0	-- base 10 floating point literal
65971.333333	-- base 10 floating point literal
65_971.333_333	-- equivalent to preceding literal
8#43.6#e+4	-- base 8 (octal) floating point literal
43.6E-4	-- base 10 floating point literal

A character literal is a single ASCII character enclosed in single quotes (examples were given in Chapter 3). A string literal is a sequence of ASCII characters enclosed in quote marks ("). An occurrence of a quote mark within a string literal is indicated by doubling the quote.

```
"ABC"
"~^!@#$%&*()_+;:'l,.?"
"doubled "" quote"
```

A bit string literal is a sequence of characters from the sets '0' to '9', 'a' to 'f', and 'A' to 'F', enclosed in quote marks and preceded by a single letter indicating the base (B for binary; O for octal; X for hex). Adjacent digits in a bit string literal may be separated with the underscore character; as is the case with integer and floating point literals, the underscore has no function other than to improve legibility.

B"11111111"
B"1111_1111"
X"fF"
X"c3"
O"70"

A physical literal consists of an integer or floating point literal plus an identifier denoting a scaling factor or unit of measurement. The following are examples of physical values:

15 ft
10 kohm
2.3 sec

An identifier denoting a unit of measurement is also a physical literal by itself; in that case, the integer literal 1 is understood to precede the unit identifier.

Scalar Types

A *scalar type* is a type whose values cannot be decomposed into more atomic values. The values of a scalar type can be ordered along a single scale (the *less-than* relation is the ordering scale). The scalar data types include integer types, floating point types, enumeration types, and physical types. Integer, floating point, and enumeration types were introduced in Chapter 3 and will be reviewed here.

The general form for all type declarations is:

type *identifier* **is** *type-definition* ;

For both integer types and floating point types, the type definition is a *range constraint*. A range constraint consists of the reserved word **range** and a *range specification*. A range specification can be constructed in one of two ways. The simplest form of range specification consists of two expressions and a *direction*. The direction is **to** for a range which is considered to be ascending and **downto** for a range which is considered to be descending. The following are range specifications constructed with expressions and a direction:

-2**31 **to** 2**31 - 1
7 **downto** 0

The ability to specify the direction of a range is particularly useful in conjunction with the definition of arrays (discussed later in this chapter). A designer may, for example, want to view a register as having bits arrayed from 7 down to 0 rather than as 0 to 7. The direction of a range is of interest for another reason: VHDL has looping constructs which will iterate over values in one order for an ascending range and in the reverse order for a descending range. The second way to construct a range specification is by using the predefined attribute 'Range. This attribute is explained later in this chapter.

The type definition for an enumeration type is a parenthesized list of enumeration literals, each enumeration literal being either an identifier or a character literal. Each identifier or character literal may appear only once in any given enumeration type definition, but a single identifier or character literal may be used in the definition of several different enumeration types.

```
type Two_level_logic is ('0', '1') ;
type Three_level_logic is ('0', '1', 'Z') ;
type Four_level_logic is ('X', '0', '1', 'Z') ;
```

The capability to have multiple meanings for an enumeration literal is actually part of a larger capability in VHDL called *overloading*. In general, an enumeration literal, an operator symbol (like "+" and "xor"), or the name of a subprogram may have multiple meanings, all visible in a VHDL description, provided that a reference to the overloaded item can be disambiguated on the basis of type information in the context of the reference. Overloading will be discussed in Chapter 9.

The fourth kind of scalar type is the physical type. The physical type definition specifies a range constraint, one *base unit*, and zero or more *secondary units*, each secondary unit being an integral multiple of the base unit.

```
type Resistance is range 1 to 10E9
  units
    ohm; -- the base unit
    kohm = 1000 ohm; -- secondary unit, multiple of base unit
  end units ;
```

There is one predefined physical type in VHDL:

```
-- The actual range of type Time is allowed to vary from
-- implementation to implementation, but must be
-- at least from -(2**31-1) to (2**31-1) femtoseconds.
```

```
type Time is range -(2**31-1) to (2**31-1)
  units
    fs ;                -- femtosecond
    ps   = 1000 fs ;    -- picosecond
    ns   = 1000 ps ;    -- nanosecond
    us   = 1000 ns ;    -- microsecond
    ms   = 1000 us ;    -- millisecond
    sec  = 1000 ms ;    -- second
    min  = 60 sec ;     -- minute
    hr   = 60 min ;     -- hour
  end units ;
```

The term *numeric type* includes integer types, floating point types and physical types. Numeric types are those for which the operations of addition, subtraction, multiplication, division, modulus, remainder, exponentiation, complement and absolute value are defined.

Composite Types

A composite type is one whose values can be decomposed into smaller atomic values. There are two kinds of composite types: records and arrays. Records are heterogeneous composite types; that is, the elements of a record can be of various types. A record type definition specifies one or more elements, each element having a different name and possibly a different type.

```
-- An enumeration type and an integer type that will be used in
-- defining a record type:
type Opcode is
  (Add, Add_with_carry, Sub, Sub_with_carry, Complement) ;
type Address is range 16#0000# to 16#FFFF# ;

type Instruction is record
  Opcode_field : Opcode ;
  Operand_1    : Address ;
  Operand_2    : Address ;
end record ;

type Register_bank is record
  F0 : Real ;
  F1 : Real ;
  R0 : Integer ;
  R1 : Integer ;
  A0 : Address ;
```

```
    A1 : Address ;
    IR : Instruction ;
end record ;
```

-- Names of elements with the same type can be grouped
-- in identifier lists, as in the following:

```
type Register_bank is record
    F0, F1 : Real ;
    R0, R1 : Integer ;
    A0, A1 : Address ;
    IR : Instruction ;
end record ;
```

Arrays are homogeneous composite types; that is, the elements of an array are all of the same type. Arrays may have one, two, or more dimensions. Each dimension has a type, and this type must be a *discrete type* (integer type or enumeration type). An array type definition may be *constrained* or *unconstrained*. In a constrained array type definition, the bounds of the array are specified, while in an unconstrained array type definition, the bounds are not specified. The ability to declare objects of unspecified size is particularly useful in interface lists. This makes it possible, for example, to declare an entity with a port that is a bundle of a variable number of bits; thus a single entity representing a shift register could serve as an 8-bit, 16-bit, or 32-bit register depending on the bounds of the actual array signal that was connected to the unconstrained port.

In a *constrained* array definition, the type and range of each dimension is specified by a *discrete range*. There are two forms of discrete range. The first form is simply a range specification, the same construct that is part of the range constraint in an integer or floating point type definition.

```
    type Word is array (15 downto 1) of Bit ;
    type Severity_level_stats is array (Note to Failure) of Integer;
```

The second form of discrete range is a *subtype indication*. A subtype indication consists of the name of a type followed optionally by a constraint. The following example declares two integer types, Column and Row, and then declares two two-dimensional constrained array types whose dimensions are specified with subtype indications. In type Matrix, the subtype indications are simply the names of the types Column and Row; thus the array has bounds of 1 to 24 and 1 to 80. In type Reduced_matrix, the subtype indications are the names of the types

Column and Row with further range constraints; the resulting array has bounds of 1 to 10 and 1 to 40.

> **type** Column **is range** 1 **to** 80 ;
> **type** Row **is range** 1 **to** 24 ;
> **type** Matrix **is array** (Row, Column) **of** Boolean ;
> **type** Reduced_matrix **is array**
>   (Row **range** 1 **to** 10, Column **range** 1 **to** 40) **of** Boolean ;

In an *unconstrained* array type, the type of each dimension is given, but the exact range and direction of each dimension, and hence the size of the array, is left unspecified (using the notation **range** <>).

> **type** Screen **is array**
>   (Integer **range** <>, Integer **range** <>) **of** Pixel;
>   -- type Pixel is defined elsewhere

There are two predefined unconstrained array types in VHDL: Bit_vector, whose index type ranges over the natural numbers, and String, whose index type ranges over the positive integers.

> **type** Bit_vector **is array** (Natural **range** <>) **of** Bit ;
> **type** String **is array** (Positive **range** <>) **of** Character ;

Aggregates and String Literals

Values of composite types can be given in aggregate notation. An aggregate is a parenthesized list of *element associations*, each element association specifying a value for an element of the array or record. Individual element associations are separated by commas. The following are examples of composite type definitions and aggregates of those types:

> -- Declaration of type Rational:
> **type** Rational **is record**
>   Numerator : Integer ;
>   Denominator : Integer ;
> **end record** ;
>
> -- Aggregate of type Rational:

(155, 2077)

-- Declaration of type Conversion_array:
type Clock_level **is** (Low, Rising, High, Falling) ;
type Conversion_array **is array** (Clock_level) **of** Bit ;

-- Aggregate of type Conversion_array:
('0', '1', '1', '0')

Just as there are alternate notations -- positional and named -- for an *association element* in an association list (discussed Chapter 3), so there are alternate notations for an *element association* in an aggregate. The aggregate in the above example is given in positional association since the mapping of each element association to an element of the composite type is given implicitly by the left-to-right order. In the alternate notation, named association, the correspondence is made explicit by naming the field of the record type element or the index value of the array element. Thus the following aggregates with named association are equivalent to those given in the immediately preceding example:

(Numerator => 155, Denominator => 2077)
(Denominator => 2077, Numerator => 155)

(Low => '0', Rising => '1', High => '1', Falling => '0')
(High => '1', Rising => '1', Falling => '0', Low => '0')

In a named element association, the record element name or array index value to the left of the arrow (=>) is called a *choice*. If several element associations specify the same value, then it is possible to group these several element associations into one. In this case, the vertical bar (|) is used to group those choices which are assigned the same value. Thus

(Low => '0', Rising => '1', High => '1', Falling => '0')

is equivalent to

(Low | Falling => '0', Rising | High => '1')

and

(Numerator => 1, Denominator => 1)

is equivalent to

(Numerator | Denominator => 1)

There are two shorthand ways of grouping choices. First, in an array aggregate, a choice may be a discrete range (as explained above, there are two forms of discrete range: a range specification and a subtype indication).

```
type Hex_letter is ('0', '1', '2', '3', '4', '5', '6', '7', '8', '9',
     'A', 'B', 'C', 'D', 'E', 'F') ;
type Hex_index is array (Hex_letter) of Boolean ;
constant Is_decimal_digit : Hex_index :=
     ('0' to '9' => True, 'A' to 'F' => False) ;
     -- Discrete ranges (range specifications) as choices

type Eight is range 1 to 8 ;
type Eight_digit is array (Eight) of Hex_letter ;
variable Hex_value : Eight_digit := (Eight => '0') ;
     -- Discrete range (subtype indication consisting of the
     -- name of a type without the additional range constraint)
     -- as a choice
```

Second, in either a record aggregate or an array aggregate, the reserved word **others** can be used to designate all remaining choices (that is, those choices, if any, not specified so far).

```
constant One : Rational := (others => 1) ;
constant Is_decimal_digit : Hex_index :=
     ('0' to '9' => True, others => False) ;
variable Hex_value : Eight_digit := (others => '0') ;
```

VHDL makes available a special kind of representation — the string literal — for values of one-dimensional arrays whose element type is an enumeration type containing character literals. The following example illustrates the use of a string literal to specify a value of a one-dimensional array.

```
type Four_level_logic is ('X', '0', '1', 'Z') ;
```

```
type Pair is array (1 downto 0) of Four_level_logic ;
constant Reset : Pair := "0Z" ;
  -- equivalent to (1 => '0', 0 => 'Z')
constant Clear : Pair := "Z0" ;
  -- equivalent to (1 => 'Z', 0 => '0')
```

Referencing Elements of Composites

An object of a composite type may be referenced in its entirety or by element. The simple name of an array or record is a reference to the entire array or record. An indexed name is used to reference an element of an array. An indexed name consists of the name of the object, followed by a parenthesized list of index expressions, one index expression for each dimension of the array. The type of an indexed name is the element-type of the array.

```
type Column is range 1 to 80 ;
type Row is range 1 to 24 ;
type Matrix is array (Row, Column) of Boolean ;

signal S : Matrix ;

-- References (indexed names) to elements of signal S:

  S (1, 1)         -- type is Boolean
  S (3, 14)        -- type is Boolean
```

A slice name is a reference to a contiguous subset of elements of a one-dimensional array (it is not possible to construct a slice name for a multi-dimensional array). The slice is constructed by appending a parenthesized discrete range to the name of the one-dimensional array. The type of a slice name is the same as that of the array being sliced.

```
type Byte is array (7 downto 0) of Bit ;
type Memory is array (0 to 2**16-1) of Byte ;

signal S_byte : Byte ;
signal S_memory : Memory ;

-- References (slice name) to contiguous elements of
-- array signals:
```

```
-- slice of three elements
S_byte (3 downto 1)

-- slice of 2**15 elements
S_memory (2**15-1 to 2**16-1)

-- slice of one element
S_memory (0) (0 downto 0)
```

It may seem counter-intuitive that the slice S_byte (3 **downto** 1) with three elements should have the same type as the declared array S_byte with eight elements. This is explained by the way VHDL defines a constrained array type such as Byte. A constrained array type declaration is actually defined to be a subtype (subtypes are explained below) of an anonymous unconstrained array type; the type is anonymous since it was not explicitly declared by the designer and therefore has no name. The *base type* (underlying type) of the slice S_byte (3 **downto** 1) is this anonymous unconstrained array type, as is the base type of the whole signal S_byte. The difference between the slice and the whole signal is a difference in index constraint, not a difference in base type.

A selected name is a reference to an element of a record. A selected name consists of the name of the object, followed by a dot (.), followed by the name of the element. The type of a selected name is the type of its element.

```
type Rational is record
  Numerator : Integer ;
  Denominator : Integer ;
end record ;

signal S : Rational ;

-- References (selected names) to elements of signal S:
S.Numerator        -- type is Integer
S.Denominator      -- type is Integer
```

Subtypes

A type declaration defines a domain of values that are completely different from the values in the domain defined by any other type declaration. Sometimes it is known that an object will take on values from only a subset of the values defined by the type. To capture such design restrictions, VHDL provides the notion of a subtype. A subtype declaration defines a subset of the values of a type by specifying a constraint on the type; a subtype is a type with a constraint. Since their values are drawn from the same domain, subtypes of a type are fully compatible with each other; however, an object declared to be of a particular subtype may not take on a value that lies outside the subset of values defined for the subtype even if that value is in the domain of the type.

Types and subtypes both provide the hardware designer assistance in checking design correctness. But the type and subtype mechanisms detect different kinds of design errors, and they detect the errors at different times in the development process. A designer can use types to detect incorrect data paths — wire connections that are in error. For example, it is an error to connect a signal of type Integer to a port of type Real (unless some type conversion is applied at the connection; type conversions are discussed in Chapter 7). Detecting such errors requires only a knowledge of the types of the objects that are being connected; a knowledge of the actual values passing along the data paths in simulation time is not required. As a result, errors involving type compatibility can be detected early in the development process, when a design unit is first analyzed. In contrast, a designer can use subtypes to detect incorrect assumptions about the actual values (for example their magnitude or sign) that an object may assume. Suppose a register of type Integer is connected to a signal of type Integer, and suppose, further, that the designer has assumed that the register will never hold values outside the range 0 to 127. If, in the course of simulating the design, a value of 128 is loaded into the register, the design has been based on a mistaken assumption. To detect such mistaken assumptions, the designer can declare the port on the register to be a subtype of Integer with the restricted range 0 to 127. Checking subtype compatibility involves checking the actual values taken on by an object against a subtype constraint; each time an object assumes a new value, the new value must be checked to see if it falls within the allowed range of values. Thus, in general, subtype checks cannot be performed until a design is simulated.

There are two ways a subtype declaration can constrain a type. The subtype can specify a range constraint that defines a subset of values of a scalar type; a value is in the subset if it falls within the range specified by the bounds and the direction (**to** or **downto**) of the range constraint. Or the subtype can specify an index constraint that, for each dimension of an unconstrained array type, defines a subset of the values defined by the discrete type of the dimension, and thereby determines a subset of the values of the array type.

```
-- Range constraint:
subtype Lowercase_letters is Character range 'a' to 'z' ;

-- Index constraint:
subtype Register is Bit_vector (7 downto 0) ;
```

Since the entire domain of values of a type is a subset of itself, it is also proper to speak of a subtype that imposes no constraint on the type. For this reason, we can say that every type declaration in VHDL also defines a subtype of the same name, and the word "subtype" can refer both to something declared in a subtype declaration and to something declared in a type declaration. The general form of a subtype declaration is

subtype *identifier* **is** *subtype-indication* ;

A *subtype indication* consists of the name of a type or subtype followed by an optional constraint. The optional constraint is either a range constraint, such as occurs in integer and floating point type declarations, or an index constraint, such as occurs in constrained array type declarations. A subtype indication that lacks an explicit constraint denotes a subtype with exactly the same constraint as the constraint on the subtype named in the type mark. Thus

subtype ASCII **is** Character ;

simply creates another name for Character.

A subtype indication may also contain the name of a function, immediately preceding the name of the type or subtype. The type returned by the function must be the same as the type named in the subtype indication. Furthermore, such a function must be a *resolution function*, that is, a function that meets the following conditions:

- The function has a single parameter whose type is an unconstrained array.
- The type returned by the function is the same as the element type of its single parameter.

If a subtype indication includes the name of a resolution function, then any signal declared to be of that subtype will be considered a *resolved signal* (see Chapter 5). A resolved signal may have multiple drivers, and the named function will determine the value of the resolved signal.

```
type Base_logic is ('0', '1', 'Z') ;
type Base_logic_array is array
   (Integer range <>) of Base_logic ;

-- A resolution function declaration:
function Resolve_drivers (
   P : Base_logic_array)
return Base_logic ;

-- A subtype indication containing the name
--   of a resolution function:
subtype Logic is Resolve_drivers Base_logic ;

-- S1 is not a resolved signal
signal S1 : Base_logic ;

-- S2 and S3 are resolved signals of the same base type
signal S2 : Logic ;
signal S3 : Resolve_drivers Base_logic ;
```

There are two predefined subtypes in VHDL: Positive and Natural. Their definition makes use of the VHDL attribute High.

```
subtype Positive is range 1 to Integer'High ;
subtype Natural is range 0 to Integer'High ;
```

Attributes

Certain classes of items in VHDL may have attributes. These classes are

types, subtypes
procedures, functions
signals, variables, constants
entities, architectures, configurations, packages
components
statement labels

An attribute is a named characteristic of items belonging to these classes. A particular attribute for a particular item may have a value, and if it does have a value, the value may be referenced by using an attribute name. The general form of an attribute name is

    name ' attribute_identifier

Attribute values are completely different from the values of object (signals, variables and constants); an object has only one value at any given time but may have many attributes. VHDL allows a designer to define his own attributes — the *user-defined* attributes — but there are also certain predefined attributes. User-defined attributes will be discussed in Chapter 9. This section discusses only predefined attributes related to types, subtypes, and array objects; other predefined attributes are discussed in Chapter 9 and in Appendix A.

    VHDL defines the attributes Left, Right, High and Low for all scalar subtypes.

    type Bit_position **is range** 15 **downto** 0 ;

    type Fraction **is range** -0.999999 **to** 0.999999 ;

    type Opcode **is**
      (Add, Add_with_carry, Sub, Sub_with_carry, Complement) ;

    subtype Adding_opcode **is** Opcode **range** Add **to** Add_with_carry ;

    -- Equalities illustrating predefined attributes:

    Bit_position'Left = 15
    Bit_position'Low = 0
    Fraction'Right = 0.999999
    Fraction'High = 0.999999
    Opcode'Left = Add
    Opcode'High = Complement
    Adding_opcode'Right = Add_with_carry

    The attributes Pos, Val, Succ, Pred, Leftof, and Rightof are defined for any physical subtype or any discrete subtype (a discrete type is an integer type or an enumeration type). Roughly speaking, the attribute Pos maps values of any discrete type onto integers while the attribute Val maps integers onto values of a discrete type. It is important to note that for an enumeration type T, T'Pos(T'Left)) is defined to be 0, while for an integer type T, T'Pos(T'Left)) is equal to T'Left.

    -- Equalities illustrating predefined attributes:

Opcode'Pos(Add) = 0
Opcode'Pos(Complement) = 4
Opcode'Val(2) = Sub
Opcode'Val(-1) is undefined
Opcode'Val(5) is undefined
Time'Pos(1 fs) = 1
Time'Pos(1 ps) = 1000
Bit_position'Pos(15) = 15
Bit_position'Val(15) = 15

For any physical or discrete type T,

T'Succ(X) = T'Val(T'Pos(X)+1)
T'Pred(X) = T'Val(T'Pos(X)-1)

For a physical or discrete type T, with an *ascending* range,

T'Rightof(X) = T'Succ(X)
T'Leftof(X) = T'Pred(X)

For a physical or discrete type T, with an *descending* range,

T'Rightof(X) = T'Pred(X)
T'Leftof(X) = T'Succ(X)

The attributes Left, Right, High, Low, Length, Range and Reverse_range, are defined on constrained array subtypes and array objects. If A is a constrained array subtype or array object, then the value of A'Left(N) is the same as the value of T'Left where T is the subtype of the Nth index of A. Right, High, and Low are defined analogously. A'Length(N) is equal to the number of elements in the Nth dimension of A. Range and Reverse_range yield ranges, not values. An attribute name of the form A'Range(N) or A'Reverse_range(N) is a *range specification*, which can be used in a range constraint and as a discrete range:

type Window **is array** (1 **to** 12, 1 **to** 40) **of** Pixel ;
-- used as a discrete range:
type Row_info **is array** (Window'Range(1)) **of** Boolean ;
-- used as a discrete range:
type Column_info **is array** (Window'Range(2)) **of** Boolean ;

-- used in a range constraint:
subtype Column_number **is** Integer **range** Window'Range(2) ;

Predefined Operators

VHDL provides a number of predefined operations for the construction of expressions that compute values. The predefined operators fall into four groups: arithmetic, relational, logical, and concatenation.

Group	Symbol	Function
arithmetic (binary)	+	addition
	-	subtraction
	*	multiplication
	/	division
	mod	modulus
	rem	remainder
	**	exponentiation
arithmetic (unary)	+	unary plus
	-	unary minus
	abs	absolute value
relational	=	equal
	/=	not equal
	<	less than
	>	greater than
	<=	less than or equal
	>=	greater than or equal
logical (binary)	**and**	logical and
	or	logical or
	nand	complement of **and**
	nor	complement of **or**
	xor	logical exclusive-or
logical (unary)	**not**	complement
concatenation	&	concatenation

The operands of the addition and subtraction operators must be of the same type and may be any numeric type (integer, floating point, or physical). The operand of the unary arithmetic operators may also be any numeric type. The operands of the modulus and remainder operators must be of the same type and may be any integer type. The operands of the multiplication and division operators may be both the same integer type or may be both the same floating point type. There are also three special cases involving multiplication or division of physical types:

- Any physical type may be multiplied by an integer literal or floating point literal.

- Any physical type may be divided by an integer literal or floating point literal.

- Any physical type may be divided by the same physical type.

The left operand of the exponentiation operator can be any integer type or floating point type; the right operand must be an integer literal (a negated integer literal is allowed as the right operand only if the left operand is of a floating point type). If one (or both) of the operands of an arithmetic operator is not a literal, then the resulting expression has the type of that operand, with one exception: the division of one physical type by the same physical type results in an expression of an integer type.

The result of an expression formed with a relational operator is always of type Boolean. The operands of a relational operator must be of the same type. The operand type of the relational operators "=" and "/=" may be any type (except a file type). The operand type of the remaining relational operators may be any scalar type or any one-dimensional array type whose element type is a discrete type (enumeration type or integer type). The following example shows the use of the "<" with one-dimensional array types.

```
type A is array (Integer range <>) of Integer ;
constant C1 : A := (2 to 1 => 1) ;  -- null array
constant C2 : A := (1 to 2 => 2) ;  -- (2, 2)
constant C3 : A := (1 => 1, 2 => 2) ;     -- (1, 2)
constant C4 : A := (1 => 1) ;      -- array of length 1

-- Given the above declarations, the following relations are true:

C1 < C4
C4 < C3
C3 < C2
```

The operands of the binary logical operators must be of the same type. The logical operators are defined for the predefined types Bit and Boolean and for one-dimensional arrays of Bit and Boolean. The resulting expression has the same type as the type of the operand. All of the binary logical operators except xor are "short-circuit" operators; that is, the right operand is evaluated only if the value of the left operand does not in itself determine the value of the binary expression. Thus, the expression

$$(X = X) \text{ or } (3/0 = 0)$$

could not generate an error condition since the right operand of the **or** will never be evaluated whereas the expression

$$(3/0 = 0) \text{ or } (X = X)$$

will always generate an error condition.

The concatenation operator is a binary operator. Either operand of the concatenation operator may be either a one-dimensional array type or the element type of a one-dimensional array type; the operands must be of the same type or one operand must be the element type of the other operand. The declarations of constants S3, S4, S5, A3, A4, and A5 below illustrate the use of the concatenation operator.

```
constant S1 : String := "ABC" ;
constant S2 : String := "DEF" ;
constant S3 : String := S1 & S2 ;   -- "ABCDEF"
constant C1 : Character := 'X' ;
constant C2 : Character := 'Y' ;
constant S4 : String := S3 & C1 ;   -- "ABCDEFX"
constant S5 : String := C1 & C2 ;   -- "XY"

type Arr is array (Integer range <>) of Boolean ;
constant A1 : Arr := (1 => True, 2 => False) ;
constant A2 : Arr := (1 => False, 2 => True) ;
constant A3 : Arr := A1 & A2; -- (True, False, False, True)
constant B1 : Boolean := False ;
constant B2 : Boolean := True ;
constant A4 : Arr := A3 & B1; -- (True,False,False,True,False)
constant A5 : Arr := B1 & B2; -- (False, True)
```

Chapter 5
Behavioral Description

This chapter explores those aspects of VHDL which relate to the behavioral description of a discrete system. The chapter is followed by a chapter on the structural description of digital devices and a chapter on system design where the description of large-scale designs is explored.

This chapter is divided into three main sections. The first section looks at the process statement, which is the primary behavioral construct in VHDL. The second section looks at the sequential nature of behavioral modeling in VHDL. This section describes the executable statements which can occur within a process statement. The final section deals with the concurrent aspects of behavioral modeling in VHDL and explores how the local sequential nature is integrated into the overall concurrency of the execution model.

Process Statements

In VHDL there are two levels at which the designer must define the behavior of a discrete system: the sequential level and the concurrent level. The sequential level involves programming the behavior of each process which will be used in the model. The concurrent level involves defining the relationship between these processes, with particular attention to the communication between them. The *process statement* (or process) is the construct which bridges these two levels.

The process statement is a concurrent statement which defines a specific behavior to be executed when that process becomes active. It has the basic form:

    process_label :
        process
            declarations
        begin
            statements
        end process;

The *process_label* defines a named label for the process; this label is optional. The *declarations* section defines the local data environment needed by the process. The *statements* section of the process is the sequential program which defines the behavior of the process.

Each process statement defines a specific action, or behavior, to be performed when the value of one of its sensitivity signals changes. This action is defined by the sequentially ordered execution statements in the process. The behavior of the process is a function of the signals read by the process. The results of the behavior are applied to the output signals and may be read by other processes. This behavior is performed each time new information is received on a sensitivity signal of the process. When the final statement of the process has been executed, execution returns to the first statement of the process and continues. Execution may be suspended using a wait statement (discussed in the next section). A process which is being executed are said to be *active*; otherwise the process is said to be *suspended*. Every process in the model may be active at any given point in time and all active processes are executed concurrently with respect to simulation time.

Process statements which do not assign values to signals are said to be *passive*. A passive process has no outputs (because it does not assign to any signals) and, therefore, cannot cause the activation of another process. Passive processes can be placed within the entity declaration of a design entity.

The Wait Statement: Activation and Suspension

As discussed above, a process is said to be in one of two states at any given time during simulation: active or suspended. The change between these states is controlled by a single construct called the wait statement. When a wait statement is executed inside a process the process suspends and the conditions for its reactivation are set. There are three different kinds of conditions: timeout, condition, and signal sensitivity. These kinds of conditions can be mixed together in the wait statement, or defaults may be used. The following forms show the three forms of the wait statement plus the default form.

> **wait**;
> **wait on** *signal-list*;
> **wait until** *condition*;
> **wait for** *time-expression*;

The first form is a wait statement which causes a process to suspend indefinitely. After the execution of this form of the wait statement, the process will not be activated again during the simulation. The second form provides a list of signals which will become the sensitivity channels of the process. Whenever an event occurs on a signal to which the process is sensitive, the process is reactivated. The third form specifies a condition which must be true before the process can be resumed. The final form supplies a maximum delay for the process to be suspended; when the time delay expires, the process is reactivated. These conditions may be mixed together to form more complex conditions. For example, consider the following behavioral description of an AND gate controlled by an enable line.

> And_Process:
> **process**
> **begin**
>   **wait on** A, B **until** Enable = '1';
>   T <= **transport** A **and** B;
> **end process**;

The statement after the wait statement is a signal assignment statement which is discussed later on in this chapter. The process will reactivate whenever there is an event on A, B or Enable but only if Enable is equal to '1', otherwise the process remains suspended.

If a process is always sensitive to one set of signals, it is possible to designate the sensitivity signals of a process using a *sensitivity list.* If this is done, it is illegal to include any wait statements in the process (or in subprograms called by the process). For example, the following process could be used to implement a simple OR gate:

> Or_Process:
> **process**(In1, In2)
> **begin**
>   Output <= In1 **or** In2;
> **end process**;

The parentheses after **process** indicate that the process will be sensitive to the signals In1 and In2. This process is equivalent to the following process:

```
Or_Process:
  process
  begin
    Output <= In1 or In2;
    wait on In1, In2;
  end process;
```

Behavioral Modeling - Sequential View

Declarations

The declarations section is used to define the local data environment of a process statement. The language allows the following declarations in this context:

> subprogram body
> variable declaration
> attribute declaration
> attribute specification
> *or*
> any of the basic declarations described in Chapter 3

Most of the constructs are discussed elsewhere in the book, so this section will focus on those which are peculiar to processes and subprograms, namely, variable declarations.

Variables are data objects which hold values specific to a single process. The declaration of a variable has the form:

> **variable** *identifier_list* : *subtype_indication*
> := *initial_value*;

A variable is declared in the declarative region of a process or subprogram. For example, the following process has a variable declaration for the variable count. The process also assigns to the variable. Variable assignments are discussed later on in the chapter.

```
P1 :
  process
    variable count : Integer := 0;
  begin
    count := count + 1;
    wait;
  end process;
```

NOTE : In this process, and throughout the text, a wait statement with no sensitivity clause, condition clause, or timeout clause is used. This will cause the process to suspend for the duration of the simulation. This type of wait statement is not generally used in practice, but is useful in examples which are used to illustrate specific points.

Sequential Assignment

The execution of a process statement may involve assigning values to variables and/or signals.

Many programming languages include global variables, meaning variables which are visible to any number of modules used in the program. VHDL does **not** allow global variables and it is important to understand the reason for this. As stated above, VHDL is primarily a controlled parallel language: controlled because the asynchronicity of the model is tempered by the timing model and parallel because each process in the model can be executing at the same time. Therefore, there is no guarantee that a given process will execute before any other process during a single simulation cycle. If one process assigned to a global variable and another process read the variable, the behavior of the system would be indeterminate. For example consider the following processes, which assumes a global variable V:

```
-- THIS VHDL IS NOT LEGAL

variable V : Integer := 0;

Process_One:
  process
  begin
    V := V + 1;
    wait;
  end process;
```

```
Process_Two:
  process
    variable X : Integer;
  begin
    X := V;
    wait;
  end process;
```

If Process_One executes before Process_Two, then X would get the value V + 1, or 1. However, if Process_Two executes before Process_One, X would get the initial value of V, or 0. There is therefore a conflict between updating a global variable and reading a global variable. Such a conflict should not be built into hardware, thus global variables have been excluded from VHDL.

Signal Assignment

The signal assignment statement is, perhaps, the most important behavioral statement in the language. Signals are used to transmit information between processes. The mechanisms of this transmission are discussed later on in the chapter. This section examines the form of the signal assignment statement *within* a process and later sections will examine the role of signals *outside* the process.

A signal is comprised of a current value and a projected output waveform. The current value element always holds the value of the signal which is read by any given process. If a signal is not a resolved signal, then there is a single process statement which assigns values to the signal. This process is called the *source* for the signal. The source of a signal may also be a connection to an **out, inout,** or **buffer** port in a child component. Component connections are discussed in the next chapter. The source of the signal may assign a number of values to the signal at a given point in time and designate a delay for each of these values. The current value of the signal is assigned a new value when the delay for the new value has expired. The collection of values and their delays is called the projected output waveform. It is possible for an element of the waveform to change or be removed before the expiration of the element's delay, causing the value to never be applied to the current value of the signal, hence *projected*. The specifics of the nature of the projected output waveform are discussed in the section of this chapter which deals with the concurrent nature of VHDL behavior. The goal of this section is to examine how a value and delay are supplied by the source process.

The simplest form of a signal assignment statement is:

signal-name <= *value*;

This statement assigns *value* to the current value of the signal at the beginning of the next simulation cycle. The left hand side of the statement is referred to as the *target* of the assignment. When no explicit delay is given in the assignment statement a *delta* delay is implied. The *value* element of the assignment can be any expression which yields the scheduled new value of the signal; e.g. a function call, an arithmetic expression, or a direct assignment of a value.

It is possible to associate the assigned value with a delay. The delay is given relative to the current time of the simulation. For example, if the current simulation time is 10 ns, then an assignment giving a delay of 2 ns will occur at 12 ns in the simulation. The signal assignment with a delay has the form:

signal-name <= *value* **after** *time-expression*;

The *time-expression* can be any expression which yields a value of the predefined type TIME.

VHDL includes two delay types for controlling the effect of the assignment on the projected output waveform of the signal: inertial and transport. Inertial delay is used to represent components which require the value on inputs to persist for a given time before the component responds. Flip-flops, where spikes on the inputs might be ignored, are a good example of this. Transport delay is similar to wire delay; *i.e.* the output always occurs no matter what the time duration of its signal. In VHDL, the delay type is always indicated in the assignment to the signal. The default delay type in VHDL is inertial. The user may indicate that the assignment should use transport delay in the following way:

signal-name <= **transport** *value* **after** *time-expression*;

The details of the effects of transport and inertial delays are explored later in the chapter.

When a number of values need to be assigned to a signal during one activation of its driving process, the following form of the signal assignment may be used:

signal-name <=
    value **after** *time-expression*,

> *value* **after** *time-expression,*
> (...)
> *value* **after** *time-expression*;

For example, the following process might be used to implement a simple clock.

```
Clock_Process:
  process
  begin
    Clock <=
            '1' after 1 ms,
            '0' after 2 ms,
            '1' after 3 ms,
            '0' after 4 ms,
            '1' after 5 ms;
    wait;
  end process;
```

One of the most common errors made when writing VHDL processes is to forget the implications of the inherent delay in a signal assignment which does not specify a time clause. It should be remembered that signals are never assigned to directly. There is always at least a delta delay between the execution of an assignment and the updating of the current value of the signal. As an example, consider the following process:

```
signal A : Bit := '0';
signal B : Bit := '1';

P1 :
  process
  begin
    B <= A;
    assert (B = A) report "B does not equal A" severity error;
    wait on B;
  end process;
```

The signal assignment statement will put the current value of A into the projected output waveform of B after a delta delay (one simulation cycle.) At the start of the next simulation cycle (*i.e.* after executing the wait statement) the value will be moved from the projected output waveform of B into the current value of B. An assertion statement provides a means for printing a message if a condition is false; the syntax and semantics of the assertion statement are discussed later in the chapter. Therefore, the condition of the assertion statement will be FALSE because when it is executed the value of A has not been moved into the current value of B. The assertion statement in the following process will, however, evaluate to TRUE and no assertion message will be reported.

```
signal A : Bit := '0';
signal B : Bit := '1';

P1 :
  process
  begin
   B <= A;
   wait on B;
   assert (B = A) report "B does not equal A" severity error;
  end process;
```

Signal Drivers

The values which are traveling over a data pathway at a given time are contained in the *driver* of the signal defining that data pathway. A driver is a collection of value/time pairs which are referred to as *transactions*. A transaction can be represented by a tuple of the form (value, time); e.g. if the value 1 is generated on the data pathway and the pathway will be updated after another 2 ns, the transaction would be designated as (1, 2 ns). Note that the time element is relative to the current simulation time so that, in this example, after one more ns this transaction would be (1, 1 ns).

A driver for a signal is defined by the process which assigns values to the signal. The driver is a special case of a *source* for the signal; the source is a process which assigns to the signal or any connection of the signal to an **out, inout,** or **buffer** port. Every process which assigns to a signal creates a source for that signal. A signal may have only one source unless the signal is a resolved signal, meaning a resolution function is associated with the signal to *resolve* the sources into a single transaction (resolved signals are discussed later on in this chapter).

For example, the following architecture contains a single source for the signal S:

```
architecture A of E is
  signal S : Integer;
begin
  process
  begin
    S <= transport 1 after 1 ns;
    S <= transport 2 after 2 ns;
    wait;
  end process;
end A;
```

Even though the process has two assignments to the signal S, there is only one process which is assigning to S. On the other hand, the following architecture contains two sources for S and, since S is not a resolved signal, is illegal.

```
-- THIS ARCHITECTURE CONTAINS AN ERROR
architecture A of E is
  signal S : Integer;
begin
  process
  begin
    S <= transport 1 after 1 ns;
    wait;
  end process;
  process
  begin
    S <= transport 2 after 2 ns;
    wait;
  end process;
end A;
```

The driver of a signal gets its initial value from the default value on the declaration which is visible to the source process. If the signal declaration and a source process for that signal are in the same architecture (or the signal is a global signal), the initial value of the driver is taken from the default value on the signal declaration. If the source process is in another component, the initial value of the driver is taken from the port declaration through which the signal is passed. This can lead to confusion in initializing models, particularly when a process statement is extracted from an architecture and placed in a child component.

For example consider an architecture which performs the NAND function on two bit values. The entity and architecture might look like:

```
entity NAND_GATE is
  port (
    A, B : in Bit;
    Y : out Bit);
end NAND_GATE;

architecture Behavior of NAND_GATE is
  signal Pass_Bit : Bit := '1';
begin
  Pass_Bit <= transport A and B;
  Y <= transport not Pass_Bit;
end Behavior;
```

At the beginning of simulation the initial value of Pass_Bit will be '1', as expected.

Suppose the architecture was to be rewritten to instantiate an AND gate in the following way:

```
entity AND_GATE is
  port (
    A, B : in Bit;
    Y : out Bit := '0');
end AND_GATE;

architecture Structure of NAND_GATE is
  signal Pass_Bit : Bit := '1';
  component AND_GATE
    port (A, B : in Bit; Y : out Bit);
begin
  And_Phase : AND_Gate
    port map (A, B, Pass_Bit);
  Y <= transport not Pass_Bit;
end Structure;
```

In this case, the initial value of Pass_Bit is taken from the default value on the **out** port of the AND_GATE; namely '0'. For a complete discussion of default values on ports see Chapter 6.

Delay in Signal Assignments

There are two types of delay that can be applied when assigning a time/value pair into the driver of a signal: transport and inertial. Transport delay is analogous to the delay incurred by passing a current through a wire. Inertial delay is used for devices that do not respond unless a value on its input persists for the given amount of time. Inertial delay is useful in modeling devices that ignore spikes on their inputs.

The effect of transport delay on the driver of a signal is fairly straightforward. If the delay time implies that the value be scheduled after all the other transactions in the driver, then a new waveform element is added to the end of the driver. If the delay time will schedule a transaction before other transactions in the driver, those later transactions are discarded. For instance, consider the following process:

```
signal S : Integer := 0;

P1 :
  process
  begin
    S <= transport 1 after 1 ns;
    S <= transport 2 after 2 ns;
    wait;
  end process;
```

If no waveform elements are on the driver of signal S before execution of this process, the effect of the execution will be to place two time/value pairs in the driver for S; namely, (1, 1 ns) and (2, 2 ns), respectively. The effects on the driver are shown in Figure 5.1.

Driver of signal S:

S ——[(1,1 ns)]

S <=transport 1 after 1 ns;

S <=transport 2 after 2 ns;

Figure 5.1.

If we reverse the statements, however, we get the following process:

signal S : Integer := 0;

P1 :
  process
  begin
    S <= **transport** 2 **after** 2 ns;
    S <= **transport** 1 **after** 1 ns;
    wait;
  end process;

This process has a different effect. The first assignment will place a transaction (2, 2 ns) on driver for S. When the second assignment is executed, this transaction will be overwritten by the transaction (1, 1 ns) because the new transaction is scheduled to occur before the transaction (2, 2 ns). The transaction (1, 1 ns) *overrides* the transaction (2, 2 ns) and, thereby, makes the second transaction obsolete. The effects on the driver are shown in Figure 5.2.

Driver of signal S:

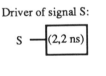

S <=transport 2 after 2 ns;

S <=transport 1 after 1 ns;

Figure 5.2.

An extension of this example will show the full effects of transport delay. Consider the following process:

signal S : Integer := 0;

P1 :
  process
  begin
    S <= **transport** 1 **after** 1 ns, 3 **after** 3 ns, 5 **after** 5 ns;
    S <= **transport** 4 **after** 4 ns;
    wait;
  end process;

The first assignment in this process adds three transactions to the driver of the signal S : (1, 1 ns), (3, 3 ns), and (5, 5 ns). The second assignment overwrites the last transaction of the previous assignment because the transaction (4, 4 ns) is scheduled before the transaction (5, 5 ns). After the execution of this process the driver of signal S will contain the transactions (1, 1 ns), (3, 3 ns) and (4, 4 ns). The effects on the driver are shown in Figure 5.3.

Driver of signal S:

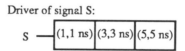

S <=transport 1 after 1 ns, 3 after 3 ns, 5 after 5 ns;

S <=transport 4 after 4 ns;

Figure 5.3.

Inertial delay is a little trickier because its effect on the driver of a signal is harder to understand at first glance. All transactions which are scheduled to occur after the delay given in the inertial assignment are discarded, as in the transport case. The difference is the effect on the transactions which are scheduled to occur **before** the new transaction. If the value of a scheduled transaction is different from the value of the new transaction, the old transaction is discarded. If the value is the same as the new transaction, the old transaction is left on the projected output waveform.

Consider the following process:

signal S : Integer := 0;

P1:
  process
  begin
    S <= 1 **after** 1 ns;
    S <= 2 **after** 2 ns;
    wait;
  end process;

NOTE: Notice that the reserved word **transport** is not included on the signal assignments. There is no reserved word **inertial** because inertial delay is the default delay type for signal assignments.

The first assignment will place the transaction (1, 1 ns) in the signal's driver. The second assignment overrides the first assignment because a new value is introduced to the driver with inertial delay. One might think of the second assignment as saying S must have the value 2 for at least 2 ns. If the first assignment were not removed, the value of S would get the value 1 after another 1 ns and then the value 2 after another 1 ns. The effects on the driver are shown in Figure 5.4.

Driver of signal S:

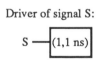

S <=1 after 1 ns;

S <=2 after 2 ns;

Figure 5.4.

Reversing the assignments has the same effect as the transport case.

```
signal S : Integer := 0;

P1:
  process
  begin
    S <= 2 after 2 ns;
    S <= 1 after 1 ns;
    wait;
  end process;
```

The second assignment overrides the first because it schedules a transaction to occur before the transaction (2, 2 ns). The effects on the driver are shown in Figure 5.5.

Driver of signal S:

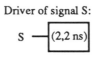

S <=2 after 2 ns;

S <=1 after 1 ns;

Figure 5.5.

A final example should show the full effect of inertial delay.

```
signal S : Integer := 0;

P1:
  process
  begin
```

```
        S <= 1 after 1 ns, 3 after 3 ns, 5 after 5 ns;
        S <= 3 after 4 ns, 4 after 5 ns;
        wait;
    end process;
```

The first assignment places the transactions (1, 1 ns), (3, 3 ns), and (5, 5 ns) on the driver for signal S. The second assignment removes the transaction (1, 1 ns) because the new transaction (3, 4 ns) has a different value component. The transaction (3, 3 ns) is retained on the projected output waveform because the value component is the same as the value component of the new transaction (3, 4 ns). The last transaction (5, 5 ns) is removed because the time component is greater than the time component of the new transaction (3, 4 ns). Finally, the second transaction from the second assignment is added to the end of the projected output waveform of signal S. The resulting projected output waveform consists of the transactions (3, 3 ns), (3, 4 ns), and (4, 5 ns). The effects on the driver are shown in Figure 5.6.

Driver of signal S:

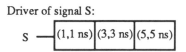

S   —[(1,1 ns)|(3,3 ns)|(5,5 ns)]

S <=1 after 1 ns, 3 after 3 ns, 5 after 5 ns;

S   —[(3,3 ns)|(3,4ns)|(4,5 ns)]

S <=3 after 4 ns, 4 after 5 ns;

Figure 5.6.

This last example highlights an important point. The semantics of the inertial delay are such that if a number of elements are given on the waveform of an inertial assignment, as in

S <= 1 after 1 ns, 3 after 3 ns, 5 after 5 ns;

the elements after the first element are not considered to be inertial assignments. This is because the language requires that the waveform elements in a signal assignment must be ascending, *i.e.* must have ascending time delays. Once the first transaction is on the driver, each subsequent transaction is guaranteed to be scheduled after all other transactions on the driver, which is the same behavior as in the transport case. That makes the above statement different from a process which has the following statements:

```
signal S : Integer := 0;

P1 :
  process
  begin
    S <= 1 after 1 ns;
    S <= 3 after 3 ns;
    S <= 5 after 5 ns;
    wait;
  end process;
```

After the execution of this process there will be a *single* transaction on the driver for S because the last assignment will override the first two assignments since its value is different. In order to get all three assignments on S, the following process must be used.

```
signal S : Integer := 0;

P1:
  process
  begin
    S <= 1 after 1 ns;
    S <= transport 3 after 3 ns;
    S <= transport 5 after 5 ns;
    wait;
  end process;
```

Variable Assignment

Variable assignment replaces the value of a variable object with a new value obtained by evaluating the expression on the left hand side of the assignment. The syntax of a variable assignment takes the form:

variable_name := expression;

A variable is declared in a process or a subprogram. When the variable is declared in a process it retains it value throughout the simulation; *i.e.* it is never reinitialized. Variables declared in subprograms are reinitialized whenever the subprogram is called. This is because processes are always executing (either active or suspended) while subprograms are only executed when they are called. If there is a wait statement in a subprogram, then the variables which are active when the process which called the subprogram suspends retain there values until the process is reactivated and the subprogram is exited.

The following process counts the events on the signal S.

```
signal S : Bit := '0';

P1 :
  process
    variable Events_on_S : Integer := 0;
  begin
    wait on S;
    Events_on_S := Events_on_S + 1;
  end process;
```

In this example, the variable Events_on_S is declared in the local data environment of the process P1. The declaration defines the variable to be of type Integer and designates that its initial value be set to 0. Whenever an event occurs on S the process will be activated and the variable Events_on_S will be updated.

It should be noted that, contrary to a signal assignment, there is no delay associated with a variable assignment. The variable object will have the new value **immediately** following the execution of the assignment statement. This means a sequential statement immediately following the variable assignment which uses the variable will see the new value of the variable.

```
P1 :
  process
    variable V : Integer := 0;
  begin
    V := 1;
    if V = 1 then
      Print("This will get printed.");
    end if;
    wait;
  end process;
```

Sequential Control

VHDL provides a number of constructs for controlling the flow of execution within a process statement. Some of these constructs provide conditional control within the process; *i.e.* statements are executed when a given condition is true or not. Others provide iterative control; statements are executed repeatedly until some condition is reached.

Conditional Control

During the execution of a process, a designer may want to select alternative collections of statements depending on some condition of the model. VHDL provides two forms for this type of conditional control, the *if statement* and the *case statement*.

The if statement is used to select one or none of a collection of statements to execute based on the truth of one or more conditions or expressions. Each condition or expression must be boolean; that is it must return either TRUE or FALSE. We can distinguish three types of conditional selection in an if statement.

The first, and most basic, form of the if statement is the selection of a collection of statements for execution based on a single condition. For instance, the following process might be used to represent a D flip-flop which assigns D to Q after a delay of 5 ns but only if the Enable port is high.

```
DFF_Process :
  process
  begin
   if Enable = '1' then
     Q <= D after 5 ns;
   end if;
   wait on D;
  end process;
```

Notice the form of the if statement:

```
if condition then
  statements
end if;
```

The second form of the if statement is the selection of one of two collections of statements for execution based upon a single condition.

For instance, the following process might be used to represent a three-state buffer. If the enable port is not enabled (equal to '0') then the buffer outputs a high impedance value ('Z'), otherwise the input is sent to the output.

```
Buffer_Process :
  process
  begin
    if Enable = '0' then
      Output <= 'Z' after Delay;
    else
      Output <= Input after Delay;
    end if;
    wait on Input;
  end process;
```

This form of the if statement takes the shape:

```
if condition then
    statements
else
    statements
end if;
```

The third, and final, form of the if statement is really a notational convenience. This form is used when there are a number of alternatives which should be executed based on a number of related or ordered conditions. The following process might be used to represent a 2-input logical AND gate.

```
And_Process:
  process
  begin
    if In1 = '0' or In2 = '0' then
      Out <= '0' after Delay;
    else
      if In1 = 'X' or In2 = 'X' then
        Out <= 'X' after Delay;
      else
        Out <= '1' after Delay;
      end if;
    end if;
    wait on In1, In2;
```

end process;

This process is legal VHDL but the nested if structure is awkward and places unwarranted emphasis on the first condition. We can rewrite this process using the third form of the if statement.

```
And_Process:
 process
 begin
   if In1 = '0' or In2 = '0' then
     Out <= '0' after Delay;
   elsif In1 = 'X' or In2 = 'X' then
     Out <= 'X' after Delay;
   else
     Out <= '1' after Delay;
   end if;
   wait on In1, In2;
 end process;
```

In this form the **elsif** branch of the if statement can be used any number of times, and with or without the final **else** branch. So assuming two uses of the **elsif** branch the third form of the if statement might look like:

```
if condition then
   statements
elsif condition then
   statements
elsif condition then
   statements
end if;
```

The *case statement* is another form of conditional control provided in VHDL. The case statement is used to select a collection of statements based on the range of values of a given expression. It is like the if statement in that a condition is used to choose among a number collections of statements. It is different, however, in that the statements are chosen based on the value of an expression. The designer gives the expression and identifies a collection of expressions for each possible value of the type of the expression. The expression must be of a discrete type or a one-dimensional array.

```
Multiplexer_Process:
 process
 begin
   case Selector is
     when "00" =>
       Out <= In0 after Delay;
     when "01" =>
       Out <= In1 after Delay;
     when "10" =>
       Out <= In2 after Delay;
     when "11" =>
       Out <= In3 after Delay;
   end case;
   wait on Selector;
 end process;
```

The **when** conditions are called *arms* of the case statement. There may be any number of these but no two arms can have the same value and every value in the range of the type must be represented. In order to facilitate the inclusion of every value in the range and to take care of types whose ranges are not finite, VHDL includes two special types of arms. The following example makes use of both of these special types in selecting statements based on an integer expression.

```
Select_Process :
 process
 begin
   case X is
     when 1 =>
       Out <= 0;
     when 2 | 3 =>
       Out <= 1;
     when others =>
       Out <= 2;
   end case;
 end process;
```

The second arm of the case statement is chosen when X evaluates to 2 *or* 3. The third arm of the case statement is chosen when X evaluates to something other than the values given in the other arms. The **when others** arm must always be the last arm of a case statement if it is included.

The general form of the case statement is:

```
case expression is
  when value =>
    statements
  when value | value =>
    statements
  when discrete-range =>
    statements
  when others =>
    statements
end case;
```

The discrete range case causes the enclosed statements to execute if the value of the expression is within the range given. The choices must be unique; *i.e.* the value of the expression must evaluate to one and only one of the arms of the case statement. There can be any number of arms in the case statement and the order can be mixed in any way except for an arm which uses the **others** keyword; if it is used, it must be the last arm of the case statement and only one such arm is allowed.

Iterative Control

Sometimes a collection of statements needs to be executed repeatedly for a specific number of times or until some condition occurs. This type of control is called iterative control because the execution iterates over the statements until some condition is met. VHDL provides iterative control in the form of loop statements or loops. There are three kinds of loop statements: the simple loop, the **for** loop and the **while** loop. The **exit** statement is a sequential statement which is closely associated with loops. The execution of an exit statement causes the loop to be exited.

A simple loop encloses a set of statements in a structure which is set up to loop forever. This structure has the form:

```
loop_label : loop
  statements
end loop loop_label;
```

The *loop_label* is optional. In the syntax of the language a loop statement can be included in the collection of statements and so it is possible to nest loop statements within other loop statements.

```
P1 :
  process
    variable A : Integer := 0;
    variable B : Integer;
  begin
    Loop1 :
    loop
      A := A + 1;
      B := 20;
      Loop2 :
      loop
        B := B - A;
      end loop Loop2;
    end loop Loop1;
    wait;
  end process;
```

This example will cause an infinite loop when executed since there is no way to get out of either loop. The simple form is useful only when used in conjunction with the **exit** statement; without it, the simulation of the model would never terminate. The exit statement has two general forms:

> **exit** *loop_label*;
> **exit** *loop_label* **when** *condition*;

The first form exits the enclosing loop with the given loop label. The second form does the same thing but only if the condition is true. In both of these forms the *loop_label* is optional; if it is not given the statement exits the immediately enclosing loop statement.

Going back to the simple loop example, we can change the process to exit when certain conditions are reached.

```
P1 :
  process
    variable A : Integer := 0;
    variable B : Integer := 1;
  begin
    Loop1 :
    loop
      A := A + 1;
      B := 20;
      Loop2 :
```

```
    loop
     if B < (A * A) then
       exit Loop2;
     end if;
      B := B - A;
    end loop Loop2;
    exit Loop1 when A > 10;
   end loop Loop1;
   wait;
 end process;
```

This looping mechanism is simple but it can lead to problems when used and does not give any indication in the loop statement itself as to how the loop will be exited. The **for** loop is used to control the exit from the loop based on iterating through some number of values in a given range. Taking the simple loop example, we can rewrite the process to reflect the intentions of the loop a little more clearly.

```
   P1 :
   process
     variable B : Integer := 1;
   begin
    Loop1:
    for A in 1 to 10 loop
      B := 20;
      Loop2:
      loop
       if B < (A * A) then
         exit Loop2;
       end if;
        B := B - A;
      end loop Loop2;
    end loop Loop1;
    wait;
   end process;
```

This is a much clearer description of the execution of the outer loop because it explicitly states the number of iterations of the enclosed statements which will be executed. The **for** loop has the general form:

```
   loop_label  :
   for loop_variable in range loop
     statements
   end loop loop_label;
```

The *loop_variable* does not need to be declared; the loop statement includes an implicit declaration for it. The *range* can be any discreet range (these are defined in Chapter 4).

The final type of loop statement is the **while** loop. This type of loop statement is used when there is a boolean condition which, when false, causes an exit from the loop. Again, looking at the example we can rewrite the process using a **while** loop for Loop2.

```
P1 :
  process
    variable B : Integer := 1;
  begin
    Loop1:
    for A in 1 to 10 loop
      B := 20;
      Loop2:
      while B >= (A * A) loop
        B := B - A;
      end loop Loop2;
    end loop Loop1;
    wait;
  end process;
```

This version of the process is clearly easier to understand and read. The **while** loop has the general form:

```
loop_label :
while condition loop
  statements
end loop loop_label;
```

One other construct which is used with loop statements is the *next statement*. This statement is used to advance control to the next iteration of a loop. The syntax for the next statement is:

next *loop-label* **when** *condition*;

The *loop-label* and **when** *condition* are both optional. When this statement is executed, control goes to the bottom of the loop identified by the label (or immediately enclosing loop if no label is given) and the loop is begun again. For a for loop, the loop index is incremented and for a while loop the condition is checked. If the *condition* is present, it is evaluated and if it is TRUE, the current loop iteration is exited and the next iteration is begun, otherwise the next statement is ignored. For instance, the P1 loop above could be rewritten as:

```
P1 :
  process
    variable B : Integer := 1;
  begin
    Loop1:
    for A in 1 to 10 loop
      B := 20;
      Loop2:
      loop
        B := B - A;
        next Loop1 when B < (A * A);
      end loop Loop2;
    end loop Loop1;
    wait;
  end process;
```

Other Sequential Statements

The Assertion Statement

The assertion statement is an interesting feature which is not found in many programming languages, much less hardware description languages. It allows the designer to encode constraints on a model within the VHDL source. The designer supplies a boolean condition which must be met in a particular device during simulation. The condition is checked during the simulation and if it is not met a message is sent to the standard output device, usually the terminal. The designer supplies the condition of the constraint, the message which will be reported if the condition becomes false, and the severity of the condition. The severity is of the type Severity_Level from package STANDARD which has the values Note, Warning, Error, and Failure. In some VHDL systems, unmet conditions of severity Error or Failure cause the simulation to terminate. The assertion statement has the following syntax:

```
assert condition
report message
```

severity *level*;

When the *condition* is FALSE the message is sent to the system output with an indication of the severity of the message and the name of the design unit in which the assertion occurred. If no message is given, the default message "Assertion violation" is reported. The default severity level is error.

> **assert** Enable /= 'X'
> **report** "Unknown value on Enable"
> **severity** Error;

This feature is very useful, especially in large-scale designs, because it allows the designer to a) directly encode specific constraints and error conditions directly into the description, b) provide useful messages which indicate the nature of the constraint or error condition, and, in some systems, c) stop the simulation when significant constraints are not met or error conditions are detected which cannot be handled by the model. For example, consider the following model of an SR flip flop:

```
entity SRFF is
  port (
    S, R : in Bit;
    Q, QBar : out Bit);
end SRFF;

architecture Constrained of SRFF is
begin
  process
    variable Internal1, Internal2 : Bit := '0';
    variable Last_State : Bit := '0';
  begin
    assert not (S = '1' and R = '1')
      report "Both S and R equal to '1'"
        severity Error;
    if S = '0' and R = '0' then
      Last_State := Last_State;
    elsif S = '0' and R = '1' then
      Last_State := '0';
    else -- S = '1' and R = '0'
      Last_State := '1';
```

```
          end if;
          Q <= Last_State after 2 ns;
          wait on R, S;
        end process;
      end SRFF;
```

If both S and R are equal to '1', the assertion is false, the message "Both S and R equal to '1'" is reported, and the simulation may be terminated (because the severity is Error).

Other examples of using assertion statements can be found in Chapter 10, particularly in the sections titled "A Device Controller" and "Setup and Hold Timing".

Procedure Calls

Procedure calls invoke procedures to be executed during a process. These procedures can be thought of as extensions to the behavior of the process statement. They serve two main purposes. First, a procedure can be written to isolate complicated sections of a process statement. This provides a further functional breakdown of the behavior and makes the process statement easier to read. Procedures are also used to encapsulate behavioral code which can be used by different processes. The behavior of the procedure is the same for all calls but the parameters may be different.

The procedure call statement has the form:

  procedure-name(association-list);

The Return Statement

The return statement is used to terminate the execution of a subprogram. For a procedure, the only legal form of the statement is

  return;

When this statement is executed, control returns to the point at which the procedure was called.

For a function, the return statement is used to return a value to the data object assigned to the function output. For instance,

```
function AND_FUNCTION(X, Y: in BIT) return BIT is
begin
  if X = '1' and Y = '1' then
    return '1';
  else
    return '0';
  end if;
end AND_FUNCTION;
```

This function returns the bit value '1' if both inputs are '1', otherwise it returns '0'.

The Null Statement

It is sometimes desirable to explicitly define conditions under which no action is to be taken. The null statement is used for this purpose. For example,

```
procedure ModTwo(X : inout Integer) is
begin
  case X is
    when  0 | 1 => null;
    when others => X := X mod 2;
  end case;
end ModTwo;
```

Behavioral Modeling - Concurrent View

The previous sections examined the local sequential nature of the process statement. The design of a discrete system is largely a matter of integrating these isolated processes into a complete system or subsystem. This section describes the concurrent statements in VHDL which can be used as shorthands for simple process statements; namely, the concurrent assertion statement and the concurrent signal assignment statement. These constructs are important in writing compact, easy to read VHDL descriptions. This is followed by a discussion of resolved signals, those signals that are assigned to by more than one process. The section, and the chapter, ends by examining a few simple examples which illustrate how processes communicate with each other.

Concurrent Statements and Equivalent Processes

There is a class of concurrent statements in VHDL which are a "shorthand" for processes which exhibit the same behavior. These statements serve to make the description easier to read and, in the case of signal assignments, to allow RTL-like expressions inside of design entities. This class of concurrent statements includes the concurrent assertion statement and concurrent signal assignment statements.

Concurrent Signal Assignment

A concurrent signal assignment statement represents an equivalent process statement which assigns to (drives) the signals given as the target of the assignment. The concurrent signal assignment is a compact way of describing a process which chooses a waveform based on a set of conditions or a control expression. The syntax of the concurrent signal assignment is similar to that of the sequential signal assignment but it is not enclosed within a process statement. The concurrent signal assignment is sensitive to the signals in its transform; *i.e.* the *longest static prefix* of every signal in the transform. The longest static prefix is a static signal or the static prefix of a composite signal. For instance, an integer signal S is itself the longest static prefix, whereas the longest static prefix of an array signal element A(I) is the entire composite signal A. A collection of concurrent signal assignments looks very much like an RTL description of a component. The concurrent signal assignment statement has the general form:

   target <= transform;

For example, consider an abstract description of a subtraction unit as described by the following design unit:

```
entity Subtractor is
  port (
    In1, In2 : in Integer;
    Out1 : out Integer);
end Subtractor;

architecture Almost_Simplest of Subtractor is
begin
  process
```

```
      begin
        Out1 <= In2 - In1 after 8 ns;
        wait on In1, In2;
      end process;
    end Almost_Simplest;
```

Using a concurrent signal assignment, the above architecture could be simplified to read:

```
    architecture Simplest of Subtractor is
    begin
      Out1 <= In2 - In1 after 8 ns;
    end Simplest;
```

The assignment is automatically executed whenever there is an event on In1 or In2.

It should be noted that, as in all concurrent statements, the textual order of two concurrent assignments does not indicate the order in which those assignments execute. For instance, the following two architectures are functionally equivalent.

```
    entity DFF is
      port (
        D : in Bit;
        Q : out Bit);

      constant Delay : Time := 5 ns;
    end DFF;

    architecture Beh1 of DFF is
    begin
      Q <= D after Delay;
      QBar <= not D after Delay;
    end Beh1;

    architecture Beh2 of DFF is
    begin
      QBar <= not D after Delay;
      Q <= D after Delay;
    end Beh1;
```

A special form of the concurrent signal assignment is the conditional signal assignment statement. In this statement, the waveform of the assignment is chosen based on a set of boolean conditions given in the assignment statement. The conditions are checked until a condition is found to be true and then the waveform associated with that condition is assigned to the target. The conditional signal assignment statement has the following form:

> *target* <= *options*
>   *waveform1* **when** *condition1* **else**
>   .
>   .
>   .
>   *waveformN-1* **when** *conditionN-1* **else**
>   *waveformN;*

The *target* is the signal(s) which is(are) being assigned to. The *options* can be either **guarded**, which is discussed in Chapter 9, or **transport** which is equivalent to the transport delay discussed earlier in the chapter. For the remainder of this section, it is assumed that **transport** is the only option used.

This behavior is similar to that of an if statement and, in fact, the equivalent process for a conditional signal assignment statement includes an if statement.

> **process** (*signals-in-transform*)
> **begin**
>   **if** *condition1* **then**
>   *target* <= *options waveform1*;
>   .
>   .
>   .
>   **elsif** *conditionN-1* **then**
>   *target* <= *options waveformN-1*;
>   **else**
>   *target* <= *options waveformN*;
>   **end if;**
> **end process;**

For example, consider the behavior of an AND gate which uses the logic states '0' and '1'. This behavior could be written with a conditional signal assignment as in the following architecture.

> **architecture** Conditional **of** AND_Gate **is**
> **begin**

```
       Y <= transport
         '1' after Delay when A = '1' and B = '1' else
         '0' after Delay;
       end Conditional;
```

This architecture is equivalent to the following architecture:

```
       architecture ConditionalEquivalent of AND_Gate is
       begin
         process (A, B)
         begin
           if A = '1' and B = '1' then
             Y <= transport '1' after Delay;
           else
             Y <= transport '0' after Delay;
           end if;
         end process;
       end ConditionalEquivalent;
```

The other kind of concurrent signal assignment is the selected signal assignment statement. Whereas the conditional signal assignment operates much like an if statement, the selected signal assignment is similar to a case statement. An expression is given and each waveform is associated with a possible value of that expression.

```
       with expression select
       target <= options
           waveform1 when choices1,

               .

               .

               .

           waveformN when choicesN;
```

The *expression* can be any expression which would be legal in a case statement and the other elements are as in the conditional assignment.

As mentioned, the equivalent process contains a case statement to represent the behavior of the assignment.

```
       process (signals-in-transform)
       begin
         case expression is
           when choices1  =>
```

```
        target <= options waveform1;
           .
           .
           .
     when choicesN =>
        target <= options waveformN;
     end case;
  end process;
```

For example, the following VHDL code defines a decoder circuit
described using a selected signal assignment statement.

```
     entity Decoder is
       port (
         Enable : in Bit;
         Sel : Bit_Vector(2 downto 0);
         DOut : out Bit_Vector(7 downto 0));
         constant Delay : Time := 5 ns;
     end Decoder;

     architecture Selected of Decoder is
     begin
       with Sel select
         DOut <=
           "00000001" after Delay when "000",
           "00000010" after Delay when "001",
           "00000100" after Delay when "010",
           "00001000" after Delay when "011",
           "00010000" after Delay when "100",
           "00100000" after Delay when "101",
           "01000000" after Delay when "110",
           "10000000" after Delay when "111",

     end Selected;
```

This architecture is equivalent to the following architecture:

```
     architecture SelectedEquivalent of Decoder is
     begin
       process(Sel)
       begin
         case Sel is
           when "000" =>
             DOut <= "00000001" after Delay;
```

```
         when "001" =>
           DOut <=  "00000010" after Delay;
         when "010" =>
           DOut <=  "00000100" after Delay;
         when "011" =>
           DOut <=  "00001000" after Delay;
         when "100" =>
           DOut <=  "00010000" after Delay;
         when "101" =>
           DOut <=  "00100000" after Delay;
         when "110" =>
           DOut <=  "01000000" after Delay;
         when "111" =>
           DOut <=  "10000000" after Delay;
       end case;
     end process;
   end SelectedEquivalent;
```

As can be seen from these examples, the concurrent signal assignment makes the behavior of certain components easier to read and more compact.

Concurrent Assertion Statement

The concurrent assertion statement represents a process statement containing an assertion statement with the same expression, report clause and severity clause given in the concurrent statement. The resulting process statement will be passive because it does not contain a signal assignment statement. It is possible, however, to include signals in the expression clause.

The concurrent assertion statement has the same syntax as the sequential assertion statement except that it can have a label and it is not enclosed inside a process. The concurrent assertion statement can be represented by an equivalent process which is sensitive to any signals which are present in the condition of the assertion. The equivalent process for a concurrent assertion statement is a passive process because the assertion does not assign to any signals. Therefore, concurrent assertions may be used in entity declarations.

As an example, consider the SR flip flop shown above in the discussion of sequential assertion statements. The design unit could be rewritten to include a concurrent assertion statement in the entity of the SR flip flop. This would make the constraint on the device evident by just looking at the interface. The new design unit is given below.

```vhdl
entity SRFF is
  port (
    S, R : in Bit;
    Q, QBar : out Bit);
begin
    SRFF_Constraint_Check:
    assert not (S = '1' and R = '1')
      report "Both S and R equal to '1'"
        severity Error;

end SRFF;

architecture Constrained of SRFF is
begin
  process
    variable Internal1, Internal2 : Bit := '0';
    variable Last_State : Bit := '0';
  begin
    if S = '0' and R = '0' then
      Last_State := Last_State;
    elsif S = '0' and R = '1' then
      Last_State := '0';
    else -- S = '1' and R = '0'
      Last_State := '1';
    end if;
    Q <= Last_State after 2 ns;
    wait on R, S;
  end process;
end SRFF;
```

The above entity is equivalent to the following entity declaration:

```vhdl
entity SRFF is
  port (
    S, R : in Bit;
    Q, QBar : out Bit);
begin
    SRFF_Constraint_Check:
    process (S, R)
    begin
      assert not (S = '1' and R = '1')
        report "Both S and R equal to '1'"
          severity Error;
    end process;
```

end SRFF;

Resolved Signals

As mentioned above, a signal normally has a single source which is *driving* the signal. VHDL allows the designer to drive a signal with more than one source, if a *resolution function* is supplied to resolve the multiple sources into a single value for the signal. A good example of when this should be used is a Wired-OR or Wired-AND, where the value to be passed to a component is the result of tying together the output of a number of gates.

A signal which has more than one source is called a *resolved* signal. A resolved signal **must** have a resolution function associated with it; *i.e.* it is an error if there are multiple sources for a signal which is not a resolved signal. If the subtype indication of the signal declaration includes a resolution function or the declaration of the subtype for the signal contains a resolution function, then the signal is a resolved signal.

A resolution function is a function which takes a one-dimensional, unconstrained array of values of the resolved type and returns a value of the resolved type. As an example, consider the following type declarations.

> **type** Bit4 **is** ('X', '0', '1', 'Z');
> **type** Bit4_Vector **is array** (Integer **range** <>) **of** Bit4;

We can declare a function which resolves signals of this type. The declaration of such a function might look like:

> **function** Wired_Or (Input: Bit4_Vector) **return** Bit4;

To declare a resolved signal using the above resolution function, two choices are available.

> **subtype** Resolved_Wire **is** Wired_Or Bit4;
> **signal** Resolved_Signal : Resolved_Wire;
>
> **signal** Resolved_Signal : Wired_Or Bit4;

In the first case, the subtype is made a resolved subtype and the signal is declared to be of that subtype. This case should be used when the same resolution function will be used for a number of signals. In the second case, the resolution function is included directly in the declaration of the signal.

The resolution function is invoked every time the value of the resolved signal is updated. If two sources are driving the signal, then the resolution function is invoked with an array of length two which contains the values of the two sources. For instance:

```
Source_1:
  process
  begin
    Resolved_Signal <= '1';
    wait;
  end process;

Source_2:
  process
  begin
    Resolved_Signal <= '0';
    wait;
  end process;
```

At the beginning of simulation these processes will execute and the signal Resolved_Signal will be updated by passing either the Bit4_Vector ('1', '0') or ('0', '1') to the resolution function and assigning the returned value '1' to the current value of Resolved_Signal. This brings up an important point: the resolution function should not depend on the order of the input values.

The body of the resolution function should handle any size of the unconstrained array. For example, the function Wired_Or is intended to return the OR function of all the inputs while disregarding any 'Z' inputs. The body of the function might look like:

```
function Wired_Or (Input: Bit4_Vector) return Bit4 is
  variable Result : Bit4 := '0';
begin
  for I in Input'Range loop
    if Input(I) = '1' then
      Result := '1';
      exit;
    elsif Input(I) = 'X' then
      Result := 'X';
```

```
        else -- Input(I) = 'Z' or '0'
          null;
        end if;
      end loop;
      return Result;
    end Wired_Or;
```

The function computes the Wired-OR and returns the result.

A Counter Element

This section will examine a very simple example of a behavioral description of an element of a counter. It will be used to show how processes communicate with each other through signals and how those signals operate.

The counter element is a simple circuit which has the following interface:

```
    entity Counter_Element is
      port (
        CIn , Clock : in Bit;
        COut, BOut : out Bit);
    end Counter_Element;
```

CIn and COut are the carry in and carry out of the element and BOut is the output bit. The body of the element is very straightforward to construct. One approach is given by the architecture:

```
    architecture Pure_Behavior of Counter_Element is
      signal DffOut : Bit := '0';
      signal ExorOut : Bit := '0';
    begin
      L1: BOut <= DffOut;
      L2: ExorOut <= DffOut xor CIn;
      L3: COut <= DffOut and CIn;
      L4: process (Clock)
      begin
        if Clock = '1' then
          DffOut <= ExorOut;
        end if;
      end process;

    end Pure_Behavior;
```

This architecture has three concurrent signal assignment statements, labeled L1, L2, and L3, and one process statement, labeled L4. The counter element is driven by the clock signal. When the clock changes to '1', a new BOut and COut are computed. During simulation the following actions are taken:

1. At the beginning of simulation, all processes are executed. This means that all the assignments and the process are executed with the initial values of the signals they are reading.

2. *First half of simulation cycle — signals are updated.* Assume there is an event on clock.

3. *Second half of simulation cycle — processes are executed.* The process labeled L4 is sensitive to Clock and is activated. If the clock is equal to '1', a transaction is scheduled for DffOut after a delta delay. If clock is not equal to '1', go back to step 2.

4. *First half of simulation cycle — signals are updated.* DffOut gets the value assigned in the last step.

5. *Second half of simulation cycle — processes are executed.* The assignments labeled L1, L2, and L3 are all sensitive to the signal DffOut, which just got a new value, so each of these assignments is executed.

Step 1 happens only at the beginning of the simulation. Steps 2 through 5 are repeated whenever there is an event on the signal Clock.

Chapter 6
Structural Description

A structural description of a piece of hardware is a description of what its subcomponents are and how the subcomponents are connected to each other. Structural description is more concrete than behavioral description; that is, the correspondence between a given portion of a structural description and a portion of the hardware is easier to see than is the correspondence between a given portion of a behavioral description and a portion of the hardware.

While the basic unit of behavioral description is the process statement, the basic unit of structural description is the component instantiation statement. Neither the process statement nor the component instantiation statement can stand by itself; each must be enclosed in an architecture body, which is a separately analyzable library unit. It is an important characteristic of VHDL that a designer can mix behavioral and structural description at any level. At a macro level, behaviorally-described components can be mixed with structurally-described components in a larger hierarchy; at a micro level, process statements and component instantiation statements can be mixed in a single architecture.

This ability to mix description modes offers the designer several advantages. First, the refinement from behavior to structure (if this is how the design process proceeds) need not proceed at the same rate for all portions of the design; thus at some stage a design may contain both abstract behavioral description for unrefined portions and a structural breakdown for portions whose refinement is known. Second, it is not necessary for a designer to simulate everything at a single, low level; portions of the design that have already been verified at a low, structural level can be replaced with behavioral versions for incorporation into larger simulations. In this way, the language allows more efficient use of simulation resources.

This chapter is divided into four sections. The first section presents the basic features of structural description in VHDL. The second section explains how regular structures can be described in VHDL. The third section explains configuration specifications. The last section treats unconnected ports and default values on ports.

Basic Features of Structural Description

Ports and their connections are the central matter of structural description. This section first explains how ports are declared in entity declarations and in component declarations and then explains how these ports can be connected by means of component instantiation statements. Three examples, an ALU, a decoder, and data bus, complete the presentation of basic features of structural description.

Ports in Entity Declarations

The external view of a hardware component is its points of connection to the outside. These points of connection could be the pins of a D connector, the pins of an IC, or the inputs and output of a NAND gate. In VHDL, the external view is given by the header of an entity declaration. The header may declare any number of ports representing points of connection. Each port declaration includes the name of the port, the direction of its data flow (the *mode* of the port), and the type of the data that flows through it.

```
entity And2 is
  port (I1 : in Bit ; I2 : in Bit ; O1 : out Bit) ;
end And2 ;
```

Ports are declared in an *interface list*; the general structure of an interface list was discussed in Chapter 3. The only object class permitted in a port list is **signal**, and as this is the default object class, it need not be specified. The default mode is **in**. Since several ports having the same mode and the same type may be grouped into a single interface element, the following is equivalent to the preceding example:

```
entity And2 is
  port (I1, I2 : Bit ; O1 : out Bit) ;
end And2 ;
```

Port Modes and Direction of Data Flow

When viewing a design at some level of abstraction above the electrical level, it is often convenient to associate a direction of data flow with a point of connection. VHDL provides for the port modes **in**, **out**, **inout**, and **buffer** for representing data flow into the component (**in**), out of the component (**out**), and both into and out of the component (**inout**, **buffer**). Assigning a particular mode to a port implies nothing about the type of data passing through the port (the type could be two-level logic, three-level logic, more abstract types like "real" or "integer", etc.). Assigning a particular mode to a port simply announces the intention of the designer to read values from the port, write values to the port, or both read values from and write values to the port. The rules of VHDL do require that values cannot be read from a port unless the mode of that port is **in**, **inout**, or **buffer** and that values cannot be written to a port unless the mode of that port is **out**, **inout**, or **buffer**. The following example demonstrates incorrect reading of an **in** port and incorrect writing to an **out** port.

```
entity Wrong_way is
  port (Not_to_be_read : out Bit ; Not_to_be_written : in Bit) ;
end Wrong_way ;

architecture Wrong_way of Wrong_way is
begin
  process
  begin
    -- The Following Line Illustrates TWO ERRORS
    Not_to_be_written <= Not_to_be_read after 5 Ns ;
    wait for 1 us ;
  end process ;
end Wrong_way ;
```

While **buffer** ports, like **inout** ports, can be both read and updated, the two modes are not fully equivalent. The differences between **buffer** and **inout** will be explained in the section on component instantiation statements.

Ports in Component Declarations

As explained in Chapter 3, ports can be declared not only in entity declarations but also in component declarations.

```
component And2
  port (I1, I2 : Bit ; O1 : out Bit) ;
end component ;
```

There are several fundamental differences between entity declarations and component declarations. First, while an entity declaration, as a separately compilable library unit, never occurs inside another library unit, a component declaration only occurs inside a library unit. A component declaration may occur inside a package declaration or inside an architecture body. Second, an entity declaration declares something that really "exists" in the design library, while a component declaration merely declares a template that does not really "exist" in the design library. This dual definition of external views gives VHDL an important flexibility. Suppose that a hardware component A consists of three subcomponents B, C, and D connected in a certain way. To describe this in VHDL, the designer would write an architecture body for A that would instantiate B, C, and D. Now an instantiation is a reference to something, and in VHDL a reference is always a reference to a defining occurrence. If an instantiation referred directly to an entity declaration, then the designer could not analyze the description for A until he had first analyzed the descriptions (entity declarations) for B, C, and D. This would place unnecessary restrictions on the way a designer is allowed to work. VHDL removes this restriction by having instantiations refer not to entity declarations but to component declarations; the associations between these component declarations and actual entities can be established separately, in a configuration declaration (configuration declarations are discussed in Chapter 7).

Component Instantiation Statements

The component instantiation statement specifies an instance of a component (a child component) occurring inside another component (a parent component). At the point of instantiation, only the external view of the child component (the names, types, and directions of its ports) is visible; signals internal to the child component are not visible. The component instantiation statement identifies the child component and specifies which ports or signals in the parent component are connected to which ports in the child component.

```
architecture Parent_body of Parent is
  component And2
    port (I1, I2 : Bit ; O1 : out Bit) ;
  end component ;
  signal S1, S2, S3 : Bit ;
begin
  Child : And2 port map (I1 => S1, I2 => S2, O1 => S3) ;
end Parent_body ;
```

The general form of the component instantiation statement is

label-identifier : *component-mark*
  generic map *association-list*
  port map *association-list* ;

The component instantiation statement *must* be preceded by a label.
(Two other statements -- the *generate statement*, explained later in this
chapter, and the *block statement*, explained in Chapter 7 -- also require
labels. These labels are referenced in *block configurations* and
component configurations, which are discussed in Chapter 7.) The
generic map associations are omitted if the corresponding component
declaration lacks generics (generics are discussed later in this chapter)
and the **port map** associations are omitted if the component declaration
lacks ports. The general structure of an association list was explained in
Chapter 3. As is generally true of associations in VHDL, either
positional or named association is possible in the association lists of a
component instantiation statement. Thus the following is equivalent to
the component instantiation statement in the preceding example:

```
Child : And2 port map (S1, S2, S3) ;
```

The *component mark*, which is either an identifier or a selected
name, must reference a component declared in a component declaration.
The component declaration need not occur in the architecture body
containing the component instantiation but it must be visible at the point
of the component instantiation. In the following example the component
declaration occurs in a package. The name of the package is made
visible by the use clause and the component declaration is made visible
by the selected name "Comp_decls.And2" (Chapter 7 contains a general
explanation of how declared items are made visible).

```
package Comp_decls is
  component And2
```

```
      port (I1, I2 : Bit ; O1 : out Bit) ;
    end component ;
  end Comp_decls ;

  use Work.Comp_decls ;
  architecture Parent_body of Parent is
    signal S1, S2, S3 : Bit ;
  begin
    Child : Comp_decls.And2
      port map (I1 => S1, I2 => S2, O1 => S3) ;
  end Parent_body ;
```

A port on a component declaration is called a *local*. In a component instantiation statement, the port association list must associate an *actual* with each local. This actual must be an object of class **signal**; an expression or an object of class **variable** or **constant** may not be associated with a local. There are two possibilities for an actual:

- The actual may be a signal declared in a signal declaration. In this case, the signal may have been declared in a package, or in the architecture containing the component instantiation statement, or in the entity declaration corresponding to the architecture containing the component instantiation statement.

- The actual may be a formal port declared in an entity declaration. In this case, the entity declaration must be the entity corresponding to the architecture containing the component instantiation statement.

VHDL imposes three kinds of restrictions on the association of an actual with a local. These restrictions are based on type, mode, and resolvability. First, VHDL requires that the type of the actual be the same as the type of the local (type mismatches can be bridged with type conversions around actuals and around locals; this is explained in Chapter 7). Second, VHDL requires that if the local is readable, then the actual must be readable, and if the local is writable, than the actual must be writable. Any signal declared in a signal declaration is both readable and writable; so, in general, a signal declared in a signal declaration may be associated with a local of any mode. However, a port of mode **in** is not writable and a port of mode **out** is not readable. It follows that an actual that is a port of mode **in** may not be associated with a local that is of mode **out** or **inout** and that an actual that is a port of mode **out** may not be associated with a local that is of mode **in** or **inout**. Third, an association with a local of mode **out** or **inout** creates a source for the actual (sources are explained in Chapter 5). It follows that an actual (either port or signal) that is not a resolved signal may not be associated with a local of mode **out** or **inout** if this would result in the actual's having more than one source. The following example illustrates erroneous port associations.

```
entity Parent is
  port (Q1 : in Bit ; Q2 : out Bit) ;
end Parent ;

architecture Parent_body of Parent is
  component Child
    port (P1 : in Bit ; P2 : out Bit ; P3 : inout Bit) ;
  end component ;
  signal S1 : Bit ;   -- S1 IS NOT A RESOLVED SIGNAL
begin
  -- The Following Line Creates a SOURCE For S1
  S1 <= not S1 after 5 Ns ;
  C1 : Child port map (
  -- The Following Line Illustrates TWO MISMATCHES
  -- OF MODE
      P1 => Q2 , P2 => Q1 ,
  -- The Following Line Would Create an
  -- ADDITIONAL SOURCE FOR S1
      P3 => S1) ;
end Parent_body ;
```

A port of mode **buffer** is like a port of mode **inout** in that they both are readable and writable; since a **buffer** port is writable, an association of an actual with a local **buffer** defines a source for the actual just as an association of an actual with a local **out** or **inout** defines a source for that actual. **Buffer** ports differ from **inout** ports in that a **buffer** port may never have more than one source (regardless of whether its subtype is resolvable or not) and that the only kind of actual that may be associated with a **buffer** port is another **buffer** port or a signal which has no other source. A **buffer** port also differs from an **inout** port in that the effective value of a **buffer** port is *always* the same as its driving value, regardless of any type conversions that may have applied to the port associations, while the effective value of an **inout** port may differ from its driving value.

The following example illustrates errors involving **buffer** ports.

```
package Types is
  function Resolve_bit (X : Bit_vector) return Bit ;
  subtype Resolved_bit is Resolve_bit Bit ;
end Types ;

use Work.Types.all ;
entity Parent is
  port (Q1 : buffer Resolved_bit ; Q2 : in Bit ; Q3 : out Bit ;
        Q4 : inout Bit) ;
```

end Parent ;

architecture Parent_body **of** Parent **is**
  component Child
    port (P1 : **buffer** Bit);
  end component ;
  signal S1 : Resolved_bit ;
begin
  -- The Following Line Creates A SOURCE FOR S1
  S1 <= **not** S1 **after** 5 Ns ;
  -- The Following Line Creates A SOURCE FOR Q1
  Q1 <= **not** Q1 **after** 5 Ns ;
  -- The Following Three Lines Illustrate
  -- MISMATCHES OF MODE
  C1 : Child **port map** (P1 => Q2) ;
  C2 : Child **port map** (P1 => Q3) ;
  C3 : Child **port map** (P1 => Q4) ;
  -- The Following Line Would ERRONEOUSLY Associate an
  -- ACTUAL BUFFER PORT That Already Has a
  -- SOURCE With a LOCAL BUFFER
  C4 : Child **port map** (P1 => Q1) ;
  -- The Following Line Would ERRONEOUSLY Associate an
  -- ACTUAL SIGNAL That Already Has a
  -- SOURCE With a LOCAL BUFFER
  C5 : Child **port map** (P1 => S1) ;
end Parent_body ;

Example: A Simple ALU

This section contains an example showing how to use the basic features of structural description in VHDL (entity declaration, component declaration, component instantiation statement). The VHDL example specifies a single stage of an ALU (Figure 6.1 is a diagram of the ALU). Inputs to the ALU stage are: four select lines S3, S2, S1, and S0; an inverted carry C1; a mode selection bit M; and the two operand bits A1 and B1. Outputs are the arithmetic or logic result F1 and the output carry C2. Logic functions are generated when M = '1', arithmetic functions when M = '0'. When M = '1', the output carry will always be '1'.

The following table shows logic functions and arithmetic functions that would be generated (on output F1) for selected values of the select lines S3 & S2 & S1 & S0

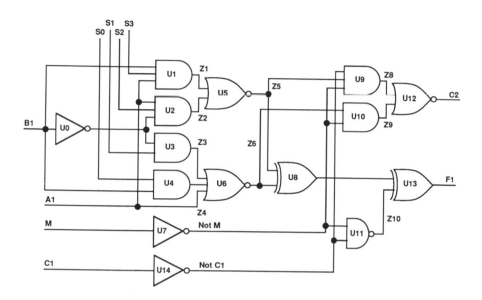

Figure 6.1. ALU Stage

if M = 1

"0000"	not A1
"0011"	'0'
"0101"	not B1
"0110"	A1 xor B1
"1010"	B1
"1011"	A1 and B1
"1100"	'1'
"1110"	A1 or B1
"1111"	A1

if M = 0

"0000"	A1 + C1
"0011"	not C1
"0110"	A1 + (1's complement B1) + C1
"1001"	A1 + B1 + C1
"1111"	A1 + 1 + C1

Here is the entity declaration for ALU_stage:

```
entity ALU_stage is
  port (S3, S2, S1, S0, A1, B1, C1, M : Bit ; C2, F1 : out Bit) ;
end ALU_stage ;
```

The following purely behavioral VHDL architecture would specify the desired behavior of the outputs C2 and F1 as direct functions of the inputs.

```
architecture Pure_behavior of Alu_stage is
begin

C2 <=
  ( (not C1) and
    ( (B1 and S3 and A1) nor (A1 and S2 and (not B1) ) )
  and (not M) )
  nor ( (not ( ( (not B1) and S1) or (S0 and B1) or A1) )
  and (not M) ) after 20 ns ;

F1 <=
  ( ( (B1 and S3 and A1) nor (A1 and S2 and (not B1) ) ) ) xor
  (not ( ( (not B1) and S1) or (S0 and B1) or A1) ) ) xor
  ( (not C1) nand (not M) ) after 20 ns ;

end Pure_behavior ;
```

This purely behavioral specification could be rewritten to represent intermediate computations as local signals; but the resulting VHDL is still behavioral since there are no component instantiation statements.

```
architecture Behavior of ALU_stage is
  signal NotC1, NotB1, NotM : Bit ;
  signal Z1, Z2, Z3, Z4, Z5, Z6, Z7, Z8, Z9, Z10 : Bit ;
  constant Delay : Time := 5 Ns ;
begin
  U0 : NotB1 <= not B1 after Delay ;
  U1 : Z1 <= B1 and S3 and A1 after Delay ;
  U2 : Z2 <= A1 and S2 and NotB1 after Delay ;
  U3 : Z3 <= NotB1 and S1 after Delay ;
  U4 : Z4 <= S0 and B1 after Delay ;
  U5 : Z5 <= Z1 nor Z2 after Delay ;
  U6 : Z6 <= not (Z3 or Z4 or A1) after Delay ;
```

```
      U7 : NotM <= not M after Delay ;
      U8 : Z7 <= Z5 xor Z6 after Delay ;
      U9 : Z8 <= NotC1 and Z5 and NotM after Delay ;
      U10 : Z9 <= Z6 and NotM after Delay ;
      U11 : Z10 <= NotC1 nand NotM after Delay ;
      U12 : C2 <= Z8 nor Z9 after Delay ;
      U13 : F1 <= Z7 xor Z10 after Delay ;
      U14 : NotC1 <= not C1 after Delay ;
   end Behavior ;
```

For the immediately preceding version there is a simple structural equivalent: each of the concurrent signal assignments using built-in operators is replaced with a component instantiation statement that instantiates a gate-level primitive (which is assumed to perform the same function as the the built-in operator).

```
   architecture Structure of ALU_stage is

      signal NotC1, NotB1, NotM : Bit ;
      signal Z1, Z2, Z3, Z4, Z5, Z6, Z7, Z8, Z9, Z10 : Bit ;

      component And2
        port (I1, I2 : Bit ; O1 : out Bit) ;
      end component ;

      component And3
        port (I1, I2, I3 : Bit ; O1 : out Bit) ;
      end component ;

      component Nor2
        port (I1, I2 : Bit ; O1 : out Bit) ;
      end component ;

      component Nor3
        port (I1, I2, I3 : Bit ; O1 : out Bit) ;
      end component ;

      component Inverter
        port (I1 : Bit ; O1 : out Bit) ;
      end component ;

      component Xor2
        port (I1, I2 : Bit ; O1 : out Bit) ;
      end component ;
```

```
component Nand2
  port (I1, I2 : Bit ; O1 : out Bit) ;
end component ;
begin
  U0 : Inverter port map (B1, NotB1) ;
  U1 : And3 port map (B1, S3, A1, Z1) ;
  U2 : And3 port map (A1, S2, NotB1, Z2) ;
  U3 : And2 port map (NotB1, S1, Z3) ;
  U4 : And2 port map (S0, B1, Z4) ;
  U5 : Nor2 port map (Z1, Z2, Z5) ;
  U6 : Nor3 port map (Z3, Z4, A1, Z6) ;
  U7 : Inverter port map (M, NotM) ;
  U8 : Xor2 port map (Z5, Z6, Z7) ;
  U9 : And3 port map (NotC1, Z5, NotM, Z8) ;
  U10 : And2 port map (Z6, NotM, Z9) ;
  U11 : Nand2 port map (NotC1, NotM, Z10) ;
  U12 : Nor2 port map (Z8, Z9, C2) ;
  U13 : Xor2 port map (Z7, Z10, F1) ;
  U14 : Inverter port map (C1, NotC1) ;
end Structure ;
```

Example: A Decoder

This section contains a further example showing how to use the basic features of structural description. We first present an abstract, behavioral specification of the decoder.

```
entity Decoder is
  port (Enable : Bit ;
        Sel : Bit_vector (2 downto 0) ;
        Dout : out Bit_vector (7 downto 0) ) ;
  constant Delay : Time := 5 Ns ;
end Decoder ;

architecture Behavior of Decoder is
begin
  with Sel select
    Dout <=
      "00000001" after Delay when "000" ,
      "00000010" after Delay when "001" ,
      "00000100" after Delay when "010" ,
      "00001000" after Delay when "011" ,
      "00010000" after Delay when "100" ,
      "00100000" after Delay when "101" ,
```

```
              "01000000" after Delay when "110" ,
              "10000000" after Delay when "111" ;
        end Behavior ;
```

Now we present the structural description of the decoder.

```
    architecture Structure of Decoder is
      component And3
        port (I1, I2, I3 : Bit ; O1 : out Bit) ;
      end component ;

      component Inverter
        port (I1 : Bit ; O1 : out Bit) ;
      end component ;

      signal Sel_bar : Bit_vector (2 downto 0) ;
    begin
      Inv_0 : Inverter
        port map (Sel(0), Sel_bar(0)) ;
      Inv_1 : Inverter
        port map (Sel(1), Sel_bar(1)) ;
      Inv_2 : Inverter
        port map (Sel(2), Sel_bar(2)) ;
      And_0 : And3
        port map (Sel_bar(0), Sel_bar(1), Sel_bar(2), Dout(0)) ;
      And_1 : And3
        port map (Sel_bar(0), Sel_bar(1), Sel    (2), Dout(1)) ;
      And_2 : And3
        port map (Sel_bar(0), Sel    (1), Sel_bar(2), Dout(2)) ;
      And_3 : And3
        port map (Sel_bar(0), Sel    (1), Sel    (2), Dout(3)) ;
      And_4 : And3
        port map (Sel    (0), Sel_bar(1), Sel_bar(2), Dout(4)) ;
      And_5 : And3
        port map (Sel    (0), Sel_bar(1), Sel    (2), Dout(5)) ;
      And_6 : And3
        port map (Sel    (0), Sel    (1), Sel_bar(2), Dout(6)) ;
      And_7 : And3
        port map (Sel    (0), Sel    (1), Sel    (2), Dout(7)) ;

    end Structure ;
```

Example: Data Bus

The two preceding examples, the ALU and the decoder, were both gate-level descriptions. As an example of higher-level structural description we will give a top-level sketch of a system consisting of several devices connected by a bus. If multiple devices are to write to a single bus, their access must be controlled by a protocol. In this example, each device that is to write to the bus is connected to a poller by one request line (port Req) and one acknowledge line (port Ack); each device is also connected to the bus (port D_bus). The device neither reads nor writes any of these ports directly; rather, each device instantiates a subcomponent Protocol which encapsulates the entire bus protocol. In this way, the common bus interface is separated from the peculiarities of each device. Each device passes the request line, acknowledge line and bus to its Protocol subcomponent. Each device has two additional connections with its Protocol subcomponent: the Write port is the means by which the device sends a write command to the protocol, and the Data port is the actual data that the device is attempting to put on the bus. Figure 6.2 shows two devices wired in the manner just described. The complete description consists of the package Bus_logic, and three design entities: entity Device_1 and its architecture Device_body; entity Device_2 and its architecture Device_body; entity Devices_on_bus and its architecture Devices_on_bus_body. The Poller and Protocol are present as component declarations; however, as their internal structure is not developed their entity declarations and architecture bodies are not given.

Package Bus_logic contains the necessary type definitions. Since there will be multiple sources for the bus, it must be a resolved signal; the type of the bus will be type Logic, a resolved subtype of type Base_logic. Type Base_logic_array, an array of the unresolved Base_logic, will be used as the type of the array of request lines and acknowledge lines on the Poller subcomponent; the arrays will be split with one wire from the array going to each device. The package also contains the component declarations for the Poller and Protocol.

```
package Bus_logic is

    type Base_logic is ('0', '1', 'Z') ;
    type Driver_array is array (Natural range <>) of Base_logic ;
    -- a resolution function
    function Resolve_drivers (P : Driver_array) return Base_logic ;
    -- resolved subtype for the bus
    subtype Logic is Resolve_drivers Base_logic ;
    constant Number_of_devices : Positive := 2 ;
    -- Base_logic_array will be used to declare the array of request
    -- and acknowledge lines on the Poller; in this example, there will
```

```
-- be two devices.
type Base_logic_array is array (1 to Number_of_devices)
   of Base_logic ;

component Protocol
   port (D_bus : inout Logic := 'Z' ; Data : Base_logic ;
        Write, Ack : Base_logic ; Req : out Base_logic) ;
end component ;

component Poller
   port (Req : Base_logic_array ;
        Ack : out Base_logic_array) ;
end component ;

end Bus_logic ;
```

In this example, the entity declarations for Device_1 and Device_2 have identical port declarations. In a more detailed example, each device would have additional ports peculiar to its own function. The three ports D_bus, Req, and Ack represent the information needed by the Protocol subcomponent which each device will instantiate.

```
use Work.Bus_logic.all ;
entity Device_1 is
   port (D_bus : inout Logic := 'Z' ;
        Req : out Base_logic ;
        Ack : in Base_logic
        -- other ports peculiar to Device_1
        ) ;
end Device_1 ;

use Work.Bus_logic.all ;
entity Device_2 is
   port (D_bus : inout Logic := 'Z' ;
        Req : out Base_logic ;
        Ack : in Base_logic
        -- other ports peculiar to Device_2
        ) ;
end Device_2 ;
```

The body of each device declares a local signal Write and a local signal Data. The signal Data holds the current value that the device wishes to put on the bus; the signal Write instructs the Protocol subcomponent to actually put the value on the bus. The device's ports D_bus, Ack, and Req are simply passed down to the Protocol subcomponent. The architecture bodies of Device_1 and Device_2 are identical in this simplified example. Additional logic, peculiar to each device, would appear in a more detailed example.

```
architecture Device_body of Device_1 is
  signal Write : Base_logic := '0' ;
  signal Data : Base_logic := '0' ;
begin
  -- instantiation of the Protocol subcomponent
  Bus_protocol : Protocol
    port map (D_bus, Data, Write, Ack, Req) ;
  -- architecture also contains logic peculiar to Device_1
end Device_body ;

architecture Device_body of Device_2 is
  signal Write : Base_logic := '0' ;
  signal Data : Base_logic := '0' ;
begin
  -- instantiation of the Protocol subcomponent
  Bus_protocol : Protocol
    port map (D_bus, Data, Write, Ack, Req) ;
  -- architecture also contains logic peculiar to Device_2
end Device_body ;
```

The top-level design unit, Devices_on_bus, instantiates the Poller and the two devices. It also contains the signal declarations for the request lines, the acknowledge lines, and the actual bus.

```
use Work.Bus_logic.all ;
entity Devices_on_bus is
end Devices_on_bus ;

architecture Devices_on_bus_body of Devices_on_bus is

  component Device_1
    port (D_bus : inout Logic := 'Z' ;
          Req : out Base_logic ;
          Ack : in Base_logic
          -- other ports peculiar to Device_1
        ) ;
  end component ;

  component Device_2
    port (D_bus : inout Logic := 'Z' ;
          Req : out Base_logic ;
```

```
        Ack : in Base_logic
        -- other ports peculiar to Device_2
        ) ;
    end component ;

    signal D_bus : Logic := 'Z' ;
    signal Req, Ack : Base_logic_array := (others => '0') ;

begin

    U1 : Device_1 port map (D_bus, Req (1), Ack (1)) ;
    U2 : Device_2 port map (D_bus, Req (2), Ack (2)) ;

    Poll_devices: Poller port map (Req, Ack) ;

end Devices_on_bus_body ;
```

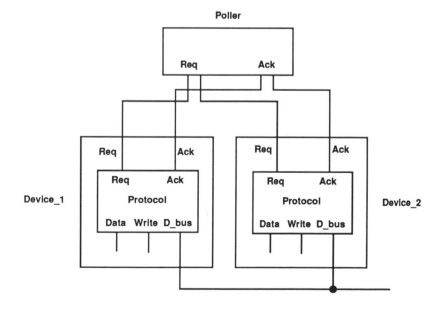

Figure 6.2. Data Bus with Two Devices

Regular Structures

Useful hardware often can be described as multiple instances of the same subcomponent, connected in some regular pattern. This section explains the generate statement, which is the means provided by VHDL to describe regular structures. This section also introduces the VHDL generic, which, when used in conjunction with the generate statement, allows the designer to specify regular structures that are general (not specific) in size.

Generate Statements

The *generate statement* consists of a *generation scheme* and a set of enclosed concurrent statements. Any VHDL concurrent statement — process statement, block statement, concurrent assertion statement, concurrent signal assignment statement, concurrent procedure call, component instantiation statement, and another generate statement — may be enclosed by a generate statement. The general form of a generate statement is

> *label-identifier* : *generation-scheme* **generate**
>   *concurrent-statements*
> **end generate** *identifier* ;

The statements portion may be empty and the closing identifier is optional. There are two kinds of generation schemes: the *if-scheme* and the *for-scheme*. Depending on the kind of generation scheme, the generate statement specifies a repetitive or conditional creation of the set of concurrent statements it contains.

A for-scheme declares a generate parameter and a discrete range defining the values that the generate parameter will take on, just as the for-scheme in the sequential loop statement (see Chapter 5) declares a loop parameter and a discrete range defining its values.

> **for** *identifier* **in** *discrete-range*

The generate parameter ("I" in the following example) acts as a constant: its value may be read but it cannot be assigned to and cannot be passed to a parameter of mode **out** or **inout**. Furthermore, its value is not defined outside the generate statement.

> **entity** Invert_8 **is**

```
    port (Inputs : Bit_vector (1 to 8) ;
      Outputs : out Bit_vector (1 to 8) ) ;
  end Invert_8 ;

  architecture Invert_8 of Invert_8 is
    component Inverter
      port (I1 : Bit ; O1 : out Bit) ;
    end component ;
  begin
    G : for I in 1 to 8 generate
      Inv : Inverter port map (Inputs(I), Outputs(I)) ;
    end generate ;
  end Invert_8 ;
```

There are no concurrent statements analogous to the sequential exit-statement and next-statement (see Chapter 5), so a generate statement with a for-scheme will always iterate over all values in the discrete range. However, it is possible for a for-scheme to specify a null range.

Regular structures often show some irregularity at the extremes. The if-generate copes with these irregularities. The following simple counter consists of T flip-flops and AND gates. But the low-bit stage and the high-bit stage are slightly different (Figure 6.3 shows the schematic).

```
  entity Counter is
    port (
      CLK : Bit ; Carry : Bit ;
      Dout : buffer Bit_vector (7 downto 0) ) ;
  end Counter ;

  architecture Counter of Counter is
    component TFF
      port (CLK : Bit ; T : Bit ; Q : buffer Bit) ;
    end component ;

    component And2
      port (I1, I2 : Bit ; O1 : out Bit) ;
    end component;

    signal S : Bit_vector (7 downto 0) ;
    signal Tied_high : Bit := '1' ;
  begin

    G1 : for I in 7 downto 0 generate
```

G2 : **if** I = 7 **generate**
  TFF_7 : TFF **port map** (CLK, S(I-1), Dout(I)) ;
end generate ;

G3 : **if** I = 0 **generate**
  TFF_0 : TFF **port map** (CLK, Tied_high, Dout(I)) ;
  S(I) <= Dout(I) ;
end generate ;

G4 : **if** I > 0 **and** I < 7 **generate**
  And_1 : And2 **port map** (S(I-1), Dout(I), S(I)) ;
  TFF_1 : TFF **port map** (CLK, S(I-1), Dout(I)) ;
end generate ;

  end generate ;
end Counter ;

Unlike the sequential if-statement, the if-generate cannot have **else** or **elsif** branches.

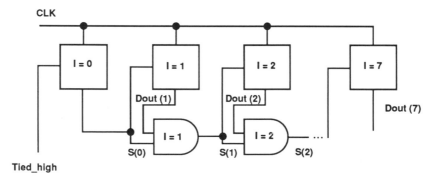

Figure 6.3. Generated Counter

An earlier example described a decoder, which exhibits a regular structure. However, the VHDL of that example did not capture this regularity. Using nested generate statements we can exhibit the basic regularity of the decoder. In the generated version, there is only one component instantiation statement in the VHDL text, and this statement describes the pattern of connections that holds for each component instance created by the nested generate statements.

```
architecture Generated_structure of Decoder is
  component And3
    port (I1, I2, I3 : Bit ; O1 : out Bit) ;
  end component ;

  component Inverter
    port (I1 : Bit ; O1 : out Bit) ;
  end component ;

  signal T : Bit_vector (5 downto 0) ;
begin

  Array_select_lines:
  for M in 2 downto 0 generate
    T(M*2+1) <= Sel(M) ;
    Inv_0 : Inverter port map (Sel(M), T(M*2) ) ;
  end generate ;

  Two_exp_2 :
  for I in 0 to 1 generate

    Two_exp_1 :
    for J in 0 to 1 generate

    Two_exp_0 :
    for K in 0 to 1 generate
      And_0 : And3 port map (T(2*2+I), T(2*1+J), T(2*0+K),
        Dout ((2**2)*I + (2**1)*J + (2**0)*K) ) ;
    end generate ;
    end generate ;
  end generate ;
end Generated_structure ;
```

The following example shows how to construct an 8-bit ALU from the description of the single ALU stage given in an earlier example. The body of the outer generate (G0) will be elaborated eight times, once for each bit-slice. The three inner generates (G1, G2 and G3) distinguish the low-order stage and the high-order stage from the other six stages; this is necessary because only the low-order stage is connected to the external carry-in and only the high-order stage is connected to the external carry-out.

```
package Reg_pkg is
  constant Size : Positive := 8 ;
  type Reg is array (Size-1 downto 0) of Bit ;
```

```vhdl
  type Bit4 is array (3 downto 0) of Bit ;
rend Reg_pkg ;

use Work.Reg_pkg.all ;

entity ALU is
  port (
    Sel : Bit4 ;
    Rega, Regb : Reg ;
    C, M : Bit ;
    Cout : out Bit ;
    Result : out Reg) ;
end ALU ;

architecture Generated_structure of ALU is
  signal Carry : Reg ;
  component ALU_stage
    port (
      S3, S2, S1, S0, A1, B1, C1, M : Bit ;
      C2, F1 : out Bit) ;
  end component ;
begin

  G0 : for I in 0 to Size-1 generate

    G1 : if I = 0 generate
      U1 : ALU_stage port map (Sel (3), Sel (2), Sel (1), Sel (0),
                  Rega (I), Regb (I), C, M,
                  Carry (I), Result (I) ) ;
    end generate ;

    G2 : if I > 0 and I < Size-1 generate
      U2 : ALU_stage port map (Sel (3), Sel (2), Sel (1), Sel (0),
                  Rega (I), Regb (I), Carry (I-1), M,
                  Carry (I), Result (I) ) ;
    end generate ;

    G3 : if I = Size-1 generate
      U3 : ALU_stage port map (Sel (3), Sel (2), Sel (1), Sel (0),
                  Rega (I), Regb (I), Carry (I-1), M,
                  Cout, Result (I) ) ;
    end generate ;

  end generate ;
end Generated_structure ;
```

Generics

Entity declarations and component declarations may declare generics in addition to ports. Generics provide a means for an instantiating (parent) component to pass values to an instantiated (child) component. Typical uses of generics are to parameterize timing, the range of subtypes, the number of instantiated subcomponents, and the size of array objects (in particular the size of ports), or simply to document physical characteristics such as temperature. In the following examples, the generics N and Size are used to constrain the size of array ports.

```
entity Decoder is
  generic (N : Positive) ;
  port (Enable : Bit ;
      Sel : Bit_vector (N-1 downto 0) ;
      Dout : out Bit_vector ((2**N)-1 downto 0) ) ;
  constant Delay : Time := 5 Ns ;
end Decoder ;

entity ALU is
  generic (Size : Positive) ;
  port (Sel : Bit_vector (3 downto 0) ;
      Rega, Regb : Bit_vector (Size-1 downto 0) ;
      C, M : Bit ; Cout : out Bit ;
      Result : out Bit_vector (Size-1 downto 0) ) ;
end ALU ;
```

Like ports, generics are declared in interface lists. The only object class permitted in a generic list is **constant**, and as this is the default object class, it need not be specified. The only mode permitted is **in**. Since a generic is a constant of mode **in**, its value can be read but it cannot be assigned to or passed to a parameter of mode **out** or **inout**. Like ports, generics appear in both entity declarations and component declarations.

```
component Decoder
  generic (N : Positive) ;
  port (Enable : Bit ;
      Sel : Bit_vector (N-1 downto 0) ;
      Dout : out Bit_vector ((2**N)-1 downto 0) ) ;
end component ;

component X
  generic (Rising_delay, Falling_delay : Time ;
    Labl : String) ;
```

    end component ;

    There are two uses of generics that are of particular relevance to structural description: using generics to control generate statements and using generics to index into arrays or define index constraints. The following example is another version of the decoder (Figure 6.4 shows the schematic). This version is generic in size; the same VHDL will describe a 1x2, 2x4, 3x8 ... decoder, depending only on the value of the generic N. Note that this example takes advantage of the fact that VHDL structure can be recursive; that is, an architecture A of an entity E can itself instantiate architecture A of E (obviously this instantiation must be inside a conditional generate or there would be no means to end the recursion).

```
entity Decoder is
  generic (N : Positive) ;
  port (Sel : Bit_vector (1 to N) ;
       Dout : out Bit_vector (1 to 2**N) ) ;
end Decoder ;

architecture Generic_structure of Decoder is

  signal Sel_bar : Bit ;
  component And2
    port (I1, I2 : Bit ; O1 : out Bit) ;
  end component ;

  component Inverter
    port (I1 : Bit ; O1 : out Bit) ;
  end component ;

  component Decoder
    generic (N : Positive) ;
    port (
      Sel : Bit_vector (1 to N) ;
      Dout : out Bit_vector (1 to 2**N) ) ;
  end component ;

begin

  Invert_select :
  Inverter port map (Sel (N), Sel_bar) ;

  Not_recursive :
  if N = 1 generate
```

```
        Dout (N) <= Sel (N) ;
        Dout (2**(N-1)+1) <= Sel_bar ;
      end generate ;

      Recursive :
      if N > 1 generate
        B1: block
          signal Temp : Bit_vector (1 to 2**(N-1) ) ;
        begin
          N_minus_1 : Decoder generic map (N - 1)
                              port map (Sel (1 to N-1), Temp) ;
          For_each_output_from_N_minus_1 :
          for I in 1 to 2**(N-1) generate
            And_each_N_minus_1_with_Sel :
              And2 port map (Temp (I), Sel (N), Dout (2*(I-1)+1) ) ;
            And_each_N_minus_1_with_Sel_bar :
              And2 port map (Temp (I), Sel_bar, Dout (2*I) ) ;
          end generate ;
        end block ;
      end generate ;

    end Generic_structure ;
```

The ALU can also be made generic in size, merely by making the constant Size a generic.

Configuration Specifications

It was explained in the section on component declarations that a component instantiation statement refers to a component declared in a component declaration, not an entity declared in an entity declaration. At some point in the design process a designer will wish to specify, for each component instance, exactly which entity declaration in which design library and which architecture of that design entity is to be selected. The *configuration specification* is the construct that allows the designer to specify the selection of entity declaration and architecture body for each component instance. The following example illustrates a simple use of the configuration specification. The example contains four library units, all assumed to be analyzed into one design library: (a) entity Inverter; (b) entity Inverter_user; (c) architecture Inverter_user of Inverter_user; and (d) architecture Inverter_body of Inverter. Entity Inverter_user contains a component declaration for a component Inv. Architecture Inverter_user contains a component instantiation statement that instantiates Inv and (preceding the component instantiation statement) a configuration specification that binds the instance of Inv to

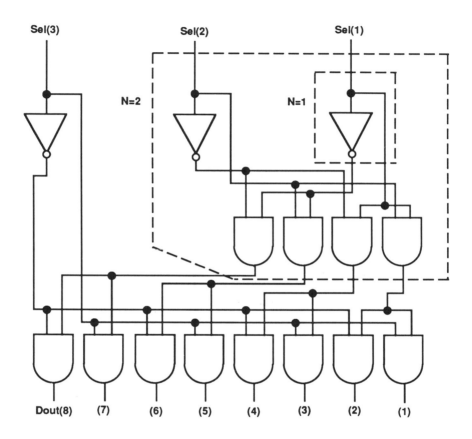

Figure 6.4. Generic Decoder

the entity Inverter and to its architecture Inverter_body. Note that it is permissible for the binding indication of a configuration specification to reference an architecture body (Inverter_body) that has not yet been analyzed.

```
entity Inverter is
  port (I1 : in Bit ; O1 : out Bit) ;
end Inverter ;

entity Inverter_user is
end Inverter_user ;

architecture Inverter_user of Inverter_user is
  signal S : Bit ;
```

```
        signal S_bar : Bit ;
        component Inv
          port (In1 : in Bit ; Out1 : out Bit) ;
        end component ;
        for U1 : Inv use entity Work.Inverter (Inverter_body)
                port map (I1 => In1, O1 => Out1) ;
      begin
        U1 : Inv port map (S, S_bar) ;
      end Inverter_user ;

      architecture Inverter_body of Inverter is
      begin
        O1 <= not I1 after 5 ns ;
      end Inverter_body ;
```

The entity declaration can differ from the component declaration in a number of ways: the entity name can differ from the component name; the entity declaration can have more ports than the component declaration; the ports on the entity declaration can have names that are different from the names of the ports on the component declaration. Clearly, the configuration specification must account for differences in names of components, names of ports, and number of ports.

The general form of the configuration specification is

> **for** *component_specification* **use** *binding_indication* ;

The *component specification* identifies which instances are configured by the configuration specification. In its simplest form, the component specification consists of an instantiation label followed by a colon and the component name; this is the form used in the example above. There are three variations on this simple form. The first variation allows a comma-separated list of instantiation labels instead of a single instantiation label; clearly all instantiation labels in the list must label instantiations of the same component.

> **for** U2, U3 : Inverter **use entity** Work.Inv1 (Inv1_body) ;

A second variation uses the reserved word **all** in place of the instantiation label; such a configuration specification applies to all instantiations of the given component.

> **for all** : And_gate **use entity** Work.And_gate1 (And_gate1) ;

A third variation uses the reserved word **others** in place of the instantiation label; such a configuration specification applies to all instantiations of the given component except for those instances whose labels appear in other (preceding) configuration specifications.

        for others : Inverter **use entity** Work.Inv2 (Inv2_body) ;

    The *binding indication* specifies the mapping between component declaration and entity declaration. The general form of a binding indication is

        entity-aspect
           generic map *association-list*
           port map *association-list*

Every binding indication will contain an *entity aspect* whose function is to name the entity declaration the designer wishes the instance to be bound to. The entity aspect consists of the reserved word **entity** followed by the name of the entity. The name of the entity is optionally followed by the simple identifier (in parentheses) of the architecture body the designer wishes the component instance to be bound to.

        entity *entity-mark* (*identifier*)

Note that the name of the entity may be qualified by the name of the library in which it resides, but that the name of the architecture body may not be so qualified. This restriction simply reflects the fact that the architecture body must be in the same library as its entity declaration. There are actually two other forms of the entity aspect. First, there may already be in a design library a *configuration declaration* that configures a subcomponent hierarchy; if this is the case, the entity aspect consists of the reserved word **configuration** followed by the name of this configuration declaration (this is discussed in Chapter 7). Second, the designer may wish to leave a component instantiation explicitly unbound; in this case the entity aspect consists of the reserved words **entity open**.

    Two instantiations of the same component may be configured so that they are bound to different entity declarations. Conversely, two instantiations of two different components may be configured so that they are bound to the same entity declaration. Figures 6.5, 6.6, and 6.7 illustrate several possible ways in which components may be bound to entities.

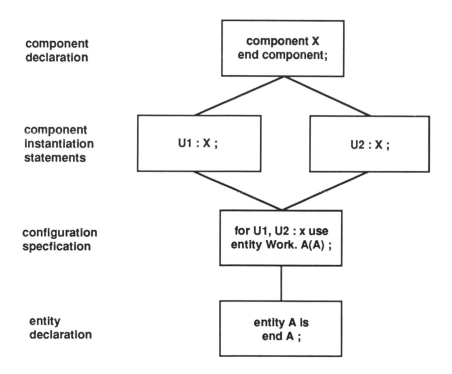

Figure 6.5. Configurations: One Component, One Entity

In addition to the entity aspect, the binding indication may also contain a generic association list and a port association list. The generic and port association lists account for differences between the generics and ports of the component declaration and the generics and ports of the entity declaration just as the entity aspect accounts for the difference between the name of the component and the name of the entity. If there is no port association list in the binding indication, then for each local port in the component declaration there must be in the entity declaration a formal port having the same name and same type as the local port. If there is no such simple match, then an explicit port association list is required. An analogous rule governs the necessity of an explicit generic association list. The syntax of the port (generic) association list in a binding indication is identical to the syntax of the port (generic) association list in the component instantiation statement. The difference is that the port or generic association list of the binding indication maps actuals to *formals* (ports or generics on an entity declaration) while the port or generic association list of a component instantiation statement maps actuals to *locals*.

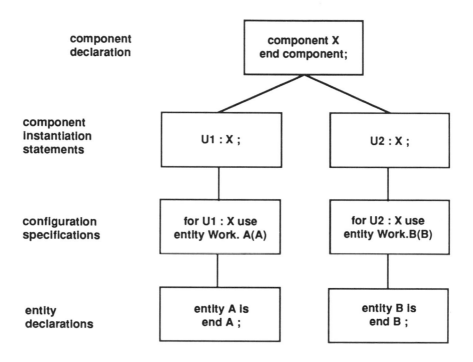

Figure 6.6. Configurations: One Component, Two Entities

The actuals in a port (or generic) association list in a binding indication are typically the local ports (or generics) of a component declaration. However, in case the entity has formals which have no corresponding locals in the component declaration, then a port (or generic) association list in a binding indication may map other signal objects to the entity formals. Consider the following example in which a signal declared in a package is associated with a formal port. In this example, the component FF is declared with only four ports. When the component instantiation statement is bound to the entity JKFF, the binding indication of the configuration specification associates the four locals of FF with four formals of JKFF and in addition associates actual signals from package Global_signals with the three additional ports of JKFF.

```
entity JKFF is
  port (CLK, PRESET, CLEAR, J, K : Bit ;
   Q, Q_bar : out Bit) ;
end JKFF ;
```

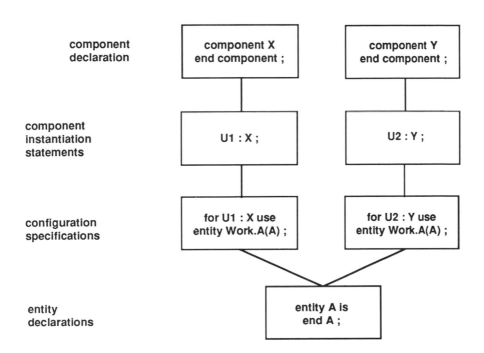

Figure 6.7. Configurations: Two Components, One Entity

```
package Global_signals is
  signal CLK : Bit ;
  signal PRESET : Bit ;
  signal CLEAR : Bit ;
end Global_signals ;

entity Design is
end Design ;

use Work.Global_signals ;
architecture Design of Design is
  signal S1, S2, S3, S4 : Bit ;
  component FF
    port (J, K : Bit ; Q, Q_bar : out Bit) ;
  end component;
  for U0 : use entity Work.JKFF
    port map (CLK => Global_signals.CLK,
            PRESET => Global_signals.PRESET,
            CLEAR => Global_signals.CLEAR,
            J => J, K => K, Q => Q, Q_bar => Q_bar) ;
```

begin
    U0 : FF **port map** (S1, S2, S3, S4) ;
    end Design ;

This section has explained the separate step of binding a
ˌcomponent instance to an entity declaration. However, it has not
attempted to explain why this separate binding is a desirable feature for
a hardware description language. To understand why separate binding is
desirable, it is necessary to understand that the binding need not be made
explicit in the architecture in which the component instantiation occurs.
Instead, the binding can occur after the architecture is written and
analyzed, in an entirely separate library unit, the *configuration
declaration.* Chapter 7 will discuss configuration declarations in detail;
for now it is sufficient to know that a single configuration declaration
can apply bindings to component instantiations occurring in architectures
that have already been analyzed into design libraries, and can thus tie
together a multi-lcvel hicrarchy of design units. The designer may write
the separate architectures without any bindings to entities, deferring the
bindings until such time as all interfaces are known, at which time a
configuration declaration can be written. The configuration declaration
thus allows a designer to delay specifying the exact interface of a
subcomponent. This has two benefits: on the one hand, it provides
support for a top-down design methodology; on the other hand, it allows
a designer to take advantage of a library of reusable subcomponents.

The separate binding mechanism of VHDL is a powerful feature,
but a designer may choose not to take advantage of the flexibility it
affords. For example, a designer may know exactly what existing design
unit he wishes to instantiate. In that case, the designer may prefer to
bind the instance right in the instantiating architecture. The
configuration specification is provided for such circumstances, as a
means by which a designer can bind an instance without having to write
a separate configuration declaration.

Default Values and Unconnected Ports

Default Values

The driver (drivers are discussed in Chapter 5) of every signal is
defined to have an implicit default value, according to the following
rules:

•   If the signal is of a scalar subtype, then the implicit default value
     of each driver is defined to be T'LEFT, where T denotes the scalar
     subtype of the signal.

- If the signal is of a composite subtype, then each scalar subelement of the composite is a signal, and each driver of each of these scalar subelements is defined to have an implicit default value by the previous rule.

Drivers of ports are defined in the same way (only ports of modes **out**, **inout**, and **buffer** can have drivers).

It is also possible to explicitly specify a default value in the declaration of the signal or port. In that case, the value of the explicit default expression overrides the implicit default value given by the above rules. The following examples illustrate signal and port declarations with explicit default expressions (the port declarations could occur in an entity declaration, in a component declaration, or in a block statement):

```
signal S : Bit := '1' ;

port (Enable : in Bit := '0' ;
      Sel : in Bit_vector (2 downto 0) := "111" ;
      Outputs : out Bit_vector (7 downto 0) := "10000000") ;
```

If a port of mode **out**, **inout**, or **buffer** does not have any sources, then a default expression on the port declaration (the port on the entity declaration, not the port on the component declaration) serves to define the initial value of the source of any actual that is connected to that port. Specifying a default expression on the declaration of a port of mode **in** serves to define the initial value of the port in case the port is unconnected (see the following section).

Initial values of drivers and signals are computed during the elaboration phase, just prior to the start of the execution phase of simulation. As part of this process, the value of a port of mode **out**, **inout** or **buffer** is propagated "upward" to contribute, as a source, to the initial value of some root signal. The value of this signal is then propagated "downward" to become the value, for reading operations, of lower-level connections of mode **in**, **inout** or **buffer**. While this concise summary appears innocuous, results may sometimes not be obvious. Consider the following example, which attempts to construct a circuit that oscillates every 10 nanoseconds by means of an inverter and buffer. The inverter and buffer were constructed to use a general three-level logic, and the ports of the inverter and buffer were declared with an explicit default expression.

```
package Three_level_logic is
  type Three is ('X', '0', '1') ;
  function Inv (Q : Three) return Three ;
end Three_level_logic ;
```

```
package body Three_level_logic is
  function Inv (Q : Three) return Three is
  begin
    case Q is
      when 'X' =>
        return 'X' ;
      when '0' =>
        return '1' ;
      when '1' =>
        return '0' ;
    end case ;
  end Inv ;
end Three_level_logic ;

use Work.Three_level_logic.all ;
entity Inverter is
  port (P1: Three := 'X' ;
        P2: out Three := 'X') ;
end Inverter ;

use Work.Three_level_logic.all ;
entity Buf is
  port (P1: Three := 'X' ;
        P2: out Three := 'X') ;
end Buf ;

architecture Inverter of Inverter is
begin
  P2 <= Inv (P1) after 5 ns ;
end Inverter ;

architecture Buf of Buf is
begin
  P2 <= P1 after 5 ns ;
end Buf ;

use Work.Three_level_logic.all ;
entity Oscillator is
end Oscillator ;

architecture Oscillator of Oscillator is

  component Inverter
    port (P1 : Three ; P2 : out Three) ;
  end component ;

  component Buf
```

```
        port (P1 : Three ; P2 : out Three) ;
    end component ;

    for all : Inverter use entity Work.Inverter (Inverter) ;
    for all : Buf use entity Work.Buf (Buf) ;

    signal S1 : Three := '0' ;
    signal S2 : Three := '1' ;

begin

    U1 : Inverter port map (S1, S2) ;
    U2 : Buf port map (S2, S1) ;

    end Oscillator ;
```

If the intention of the designer of the above circuit is for the circuit to begin oscillating at simulation startup and continue indefinitely, then the design is in error. Both signal S1 and signal S2 are connected to ports of mode **out**; hence the initial values of S1 and S2 will be taken from the values of the ports they are connected to. These ports, named P2 in each case, are declared with the default expression 'X'; hence the initial value of the driver of each P2 will be 'X'. Consequently, the initial value of the signals S1 and S2 will be 'X', despite the default expressions '0' and '1' on the signal declarations for S1 and S2. And since these values will be propagated downward, it follows that the initial value on each of the ports named P1 will be also be 'X'. But since the inversion of 'X' is 'X', the initial execution of the concurrent signal assignment statements in Inverter and Buf will not result in any change of state, and the simulation will terminate after 5 nanoseconds. A solution to this problem is change the Inv function so that the inversion of 'X' is either '0' or '1'. Alternatively, the explicit default expression on the **out** ports of the two entities could be changed to '0' or '1'.

Note that a local port of mode **out, inout,** or **buffer** may also play a role in determining the initial value of a signal with which the port has been associated. If the subtype of the local port is a resolved subtype, then the resolution function will be invoked during the upward propagation of values through the local port, and it is this resolved value that will be propagated as a source for the actual signal.

Unconnected Ports

It may be that a component is declared to have one or more ports
that are not needed by an instantiating architecture. For example, a JK
flip-flop may have been declared with inputs CLK, Preset, CLR, J, and
K and outputs Q and Q_bar, while a particular instantiation may have no
need of the Q_bar output. Nevertheless, the rules of VHDL require that
in a component instantiation there be an association for each local port.
In such cases, the designer can always create a local dummy signal and
associate this otherwise unused dummy signal with the extra port.

```
package Flip_flops is
  component port JKFF (
    CLK, PRESET, CLEAR, J, K : Bit ;
    Q, Q_bar : out Bit) ;
  end component ;
end Flip_flops ;

use Work.Flip_flops ;
architecture FF_user of FF_user is
  signal CLK, Preset, CLR, In_1, In_2, Out_1 : Bit ;
  signal Dummy : Bit ;
begin
  C : Flip_flops.JKFF
    port map (CLK, Preset, CLR, In_1, In_2, Out_1, Dummy) ;
end FF_user ;
```

While this solution will simulate correctly, it is awkward. In
particular, it is indistinguishable from a design error in which a wire was
left dangling. A better alternative is to leave the extra port explicitly
unconnected by associating with it the reserved word **open**.

```
use Work.Flip_flops ;
architecture FF_user of FF_user is
  signal CLK, Preset, CLR, In_1, In_2, Out_1 : Bit ;
begin
  C : Flip_flops.JKFF
    port map (CLK, Preset, CLR, In_1, In_2, Out_1, open) ;
end FF_user ;
```

Any port of mode **out**, **inout**, or **buffer** may be unconnected
provided its type is not an unconstrained array type. A port of mode **in**
may be unconnected only if its declaration includes an explicit default

expression.

The reserved word **open** may also be used in the port association list of a binding indication. In the following example, **open** is associated with the formal port Q_bar of entity JKFF.

```
package Flip_flops is
  component port JKFF (CLK, PRESET, CLEAR, J, K : Bit ;
    Q : out Bit) ;
  end component ;
end Flip_flops ;

entity JKFF is
  port JKFF (CLK, PRESET, CLEAR, J, K : Bit ;
    Q, Q_bar : out Bit) ;
end JKFF ;

use Work.Flip_flops ;
architecture FF_user of FF_user is
  signal CLK, Preset, CLR, In_1, In_2, Out_1 : Bit ;
  for C : Flip_flops.JKFF use entity Work.JKFF
    port map (CLK, PRESET, CLEAR, J, K, Q, open) ;
begin
  C : Flip_flops.JKFF
    port map (CLK, Preset, CLR, In_1, In_2, Out_1) ;
end FF_user ;
```

VHDL also defines a default port association list which associates each local port with a formal port of the same name (the type and mode of each local port must be the same as the type and mode of the formal port is associated with) and associates **open** with each formal port for which there is no corresponding local port. If, in the immediately preceding example, the port association list were omitted from the binding indication, then the default port association list would apply, with the same results as obtained with the explicit port association list.

Chapter 7
Large Scale Design

The previous chapters have shown how structural and behavioral descriptions may be written using VHDL. Most of the examples have been small-scale designs. However, VHDL contains capabilities that allow and support the development of large-scale designs, and designs that include arbitrary mixtures of structure and behavior — language features that support design at the system level. This chapter takes the elements of VHDL presented in previous chapters and shows how the language fits them together to support this level of design.

The first section discusses the concept of libraries of designs as a way of maintaining and organizing multiple designs or projects and built or purchased parts catalogs. The succeeding section describes VHDL's name visibility rules, which determine what a particular name means at a particular point in a design. The **use** clause and **library** clause, which between them provide access to saved libraries, are discussed as well. Next come two sections that discuss how designs are partitioned; the first of these discusses the partitioning of algorithms and structure, while the second discusses the partitioning of data. Once a design is partitioned and completed, there must be a way to pull the pieces back together to describe the configuration of the completed design; this mechanism, the *configuration declaration*, is discussed in the fifth section. The chapter concludes with a brief discussion of combining behavioral and structural descriptions in a single design, and the concept of refining a design to introduce successively more levels of detail; these ideas are discussed in detail in Chapter 8, which contains a complete example of the process of developing a small VHDL design.

Managing Shared Designs

The complexity of hardware designs has greatly increased over the past number of years. With the introduction of VLSI and VHSIC technology, it is not uncommon to find designs which place 100,000 gates or more on a single chip. Managing the various components of such a design, therefore, becomes a complex and error-prone task. Even with the decomposition aids in the VHDL language, the problems of configuration control and revision control remain. A second, but equally important, issue is the problem of importing or using libraries of previously developed designs and allowing library modification when, and only when, appropriate. VHDL addresses these issues through the use of libraries of design units.

Design Libraries and their Implementation

A library can be used to store the components of a given section of a design. A design library may contain any number of previously analyzed design units. A particular VHDL language analysis tool (an analyzer) may allow the use of any number of libraries; the specific syntax for stating this in a VHDL design is described in the section on Visibility later in this chapter.

A design library is typically represented as a file or set of files on a disk, with host operating system dependent naming and organization. In order to increase the level of portability of design descriptions, VHDL abstracts this host-dependent information out of the language. Thus, instead of referring to a design library via its host file name, a logical name is used — a name that is interpreted by the toolset and mapped into a reference to the actual design library. For example, a design library containing VHDL definitions of the parts in the TTL Data Book might be stored in a Unix subdirectory named

/usr/vhdl/std_parts/ti_ttl

or in a VMS subdirectory such as

USERDISK0:[VHDL.STD_PARTS.TI_TTL]

Within a VHDL design, either of these libraries might be referred to simply as "TI_TTL", or as "TTL", or by whatever other logical name the users or system administrators choose to ascribe to this design library. In fact, if a VHDL design is ported from Unix to VMS, and the toolset on VMS provides a means of reassigning logical names to design libraries, no system administration work or coordination of library names is required. For instance, the design on the Unix system might refer to the library as "TTL". When ported to VMS, if the VMS-based toolset provides a way to associate the logical name "TTL" with

USERDISK0:[VHDL.STD_PARTS.TI_TTL],

no changes would be required to the design library references.

This capability provides a great deal of flexibility, when combined with a tool implementation that fully supports and encourages the use of the full generality of the library mechanism. However, it is important to remember that the library mechanism is one of the places in the language where the boundary between the abstraction of the design language and the harsh reality of the host operating system is encountered. Two specific ways in which this boundary is evident are: 1) the absence of a language-defined mapping between library logical names and host library names, and 2) the absence in the language *per se* of other important notions related to libraries, such as revision numbers.

Predefined Design Libraries

Each VHDL implementation is required to support the use of two library logical names: WORK and STD. Of course, as discussed above, an implementation may and should choose to support a large number of libraries in order to increase the flexibility available to the designer.

The WORK library is a library that denotes the current working library during analysis; the results of analyzing a particular unit are placed into the WORK library for use by further analysis processes. The WORK library may be, in a particular implementation or within a particular set of management policies, either a private library or a library that is used by several people in development of a larger system.

The STD library is a library whose contents are defined by the VHDL language. The STD library contains two packages. (Packages were touched on briefly in Chapter 3, Basics, and are described in some detail in a later section of this chapter). The first, package STANDARD, is fairly simple. It defines the following things:

- The BIT, BIT_VECTOR, BOOLEAN, CHARACTER, and STRING types;
- The SEVERITY_LEVEL type for use in **assert** statements;
- The INTEGER and REAL types, together with the NATURAL and POSITIVE subtypes of INTEGER;
- Type TIME; and

- The function NOW, which returns the current simulation time.

The VHDL definition of package STANDARD is given in Appendix A of this book; all elements of the package were described in Chapters 3 and 4.

The second package in the STD library is package TEXTIO. This package contains declarations for types and subprograms that support ASCII I/O operations and the file subsystem of VHDL. The text of this package is also given in the Appendices; it contains, in addition to basic definitions, procedures to read and write objects of each of the predefined types, as well as procedures to read and write a complete line, and functions to test for end-of-line or end-of-file conditions. These aspects of the language are discussed in more detail in Chapter 9, Advanced Features.

The STD library thus represents the language interface into the underlying simulation host and simulation model. It contains a number of implementation-dependent types and functions. Typically, the implementation of the packages in STD will not appear as VHDL, but rather will be hidden within the runtime system (or simulation kernel) of the toolset's simulator. In order to improve design portability, and to force explicit declaration of the usage of nonstandard library parts, the language reference manual recommends that language implementations place only these two packages into library STD. Other implementation-specific packages should be placed into implementation-specific libraries.

An intriguing possibility is suggested by the above discussion of the STD library. One could easily imagine that a particular implementation might develop a set of VHDL package declarations, or entity declarations, whose implementations are not in fact written in VHDL, but rather are written in some other source language, or are in fact direct interfaces into the simulation kernel or runtime library. For example, a package to add two's complement BIT_VECTORs, or a multilevel logic modeling package, are obvious candidates for such an implementation — the performance benefits to be gained are clear. While implementations such as these are attractive, there is a cost paid in portability of the resulting designs, since implementation-specific packages that do not have VHDL bodies must be re-implemented for each new host.

The Use of Libraries for Revision Management

Previous versions of the VHDL language contained support for the notion of a revision of a design entity, and also had the notion of referring to a particular revision of a component part in a configuration specification. The IEEE language design has removed that capability from the language, feeling that management of revisions is more appropriately a function of the design support environment rather than of the language itself. This is both a good and a bad decision. On the one hand, a strong argument can be made that the complexities of revision management in the language were sufficient that a completely language-based solution would be very difficult to design properly. On the other hand, it is valuable to have, within a particular model and configuration, the definition of precisely which revisions of what parts were used to construct it.

Removing this capability from the language has not, however, vitiated the need for the capability itself. It has simply placed on the environment the burden for providing it. For example, one could imagine an environment in which the logical name definitions were translated by a rather complex tool or set of procedures called by the VHDL analysis tool; these procedures could decide, based on user inputs or a given list of libraries to be searched, or other user-provided criteria such as date/time stamps, which revisions of components to select for use. Although no existing VHDL system fully implements such a capability, at least one system incorporates the concept of "sublibraries" that may be selected by the user; this allows some amount of flexibility in revision management and selection. A full configuration management and revision management capability is clearly necessary for lifecycle support of large scale designs. Such configuration management systems are common practice in many aspects of engineering; at some point, they will have to be integrated with a VHDL support environment.

Visibility and the Analysis Context

Name Visibility in VHDL

When a group of people cooperates to build a large design in VHDL, or when a design is built up out of piece parts that are already designed, there are clearly potential problems with regard to naming of components. There is no guarantee that two designers will not select the same identifier to identify different objects in their respective portions of the design. A language that claims to support large-scale design must certainly allow for partitioning, or privatization, of the set of names when appropriate, and must allow for some measure of control over which names may be referenced at a particular point in the design. The purpose of this section is to describe those rules. Those readers familiar with Pascal or Ada, particularly Ada, will see concepts in this section that are reminiscent of those languages.

In many languages (both HDL's and programming languages), particularly older languages, visibility is not a particularly interesting topic: often the design, or program, is entered in a single file and any name declared before a given point may be used there. As languages become more complicated, enabling them to support a larger scale of implementation problems, this approach becomes inadequate. For this reason, modern languages such as VHDL allow for nesting of various constructs within others (like procedures in Pascal, or packages, procedures or tasks in Ada); each nested construct introduces a new *name space*. (These name spaces are called *declarative regions* in the VHDL Language Reference Manual). As implied by the term, a name space is a region within which a name may be declared without conflicting with another declaration of the same name outside of the name space; the declaration simply overrides, or "hides", it within the name space. A second important attribute of a name space is that any name declared within it is normally not visible outside of it. In some languages, this second attribute is an iron clad rule; the only way to pass information out of a name scope is by procedure parameters. However, VHDL allows for multiple independently analyzed units (entity declarations, configuration declarations, package declarations, architecture bodies, and package bodies), and allows one unit to gain access to names declared within another unit by means of the **library** and **use** clauses, which were discussed briefly in Chapter 3, Basics, and are discussed in more detail below. For this reason, more discussion of visibility is required in order to fully understand it.

The first thing to do is to identify the VHDL constructs that create new name spaces. These are:

- An entity declaration and an associated architecture body.
- A block statement.
- A process statement.
- A subprogram (procedure or function) declaration, along with its body.
- A configuration declaration.
- A record type declaration. This warrants a little more explanation. A record type declaration creates a name space in order that the person defining the record type component names need not worry about name conflicts. Thus, for example, record types NET and WIRE could each have a component named PINS.
- A loop statement. In this case, the only reason for making a loop statement create a name space is so that the loop variable, which is implicitly declared by its appearance in the loop's iteration scheme, will not be visible past the end of the loop. Note that even though this creates a name space, no other names may be declared immediately within it.
- A component declaration, a block configuration, or a component configuration. These are all constructs that are discussed later in this chapter, in the section on Specifying a Design Configuration.

A name declared within the current name space is visible from the point of its declaration until the end of the name space. All this means is that if you declare a variable within a procedure, say, or a signal within an architecture, then you can use that name as you might expect. Actually, this statement is slightly too simplistic, since if there is a construct nested within your procedure or architecture (say a block within the architecture) that declares another signal with the same name, then within that nested construct, references to that signal name will refer to the nested declaration. Thus,

```
-- a block declaration
B1: block
    -- containing a signal declaration
    signal SUM, I1, I2: Bit;
begin

    -- a nested block declaration
    B2: block
        -- This SUM hides the one in block B1,
        signal SUM: Bit;
    begin
        -- so this assigns to the SUM in B2,
        SUM <= I1 and I2 after 4 ns;
    end block;

    -- whereas this assigns to the SUM in B1.
    SUM <= I1 and I2 after 5.5 ns;

end block;
```

Even if a name is declared in a nested construct that hides an object of the same name in an outer construct, the outer name may be referenced as well. In the previous example, the statement

$$B1.SUM <= I1 \text{ and } I2 \text{ after } 4 \text{ ns;}$$

within block B2 would nonetheless have referred to the SUM declared in block B1. This is a general mechanism in the language for referring to names declared in name spaces that surround the current name space (such as procedures within procedures, loops within procedures, procedures within packages, etc.).

This idea of referring to objects by giving the name of the name space within which the object is declared, followed by the name of the object itself, derives from the notion of referring to record components in that form: "NET.PINS" means the PINS component of the NET record. The notation is further extended in VHDL to provide access to name spaces that do not surround the current name space. Thus, this "dot notation" (also known as *selected notation*) may be used to refer to a package, entity, or configuration declared within a library or to a declaration within a package declaration, in addition to the uses already mentioned. For example, to refer to a global clock signal, one might write:

PROJECT_LIB.GLOBAL_SIGNALS.CLOCK

which states that CLOCK is in package GLOBAL_SIGNALS, which is located in the library PROJECT_LIB. An object that is referenced in this way is said to be *visible by selection*.

Of course, having to type all of this each time reference to something in the package was desired would be very cumbersome; the *use clause* is designed to provide a shorthand. The use clause specifies a list of names that are to be *directly visible*. In other words, these names may, after the use clause appears, be given without the prefix. In the example above regarding the clock signal, the clause

use PROJECT_LIB.GLOBAL_SIGNALS;

would allow reference to the clock by saying "GLOBAL_SIGNALS.CLOCK", *i.e.*, GLOBAL_SIGNALS would be directly visible. The clause

use PROJECT_LIB.GLOBAL_SIGNALS.CLOCK;

would allow reference to the clock by simply saying "CLOCK". In this case, CLOCK is directly visible. This could also be achieved by saying

use PROJECT_LIB.GLOBAL_SIGNALS;
use GLOBAL_SIGNALS.CLOCK;

since after the first use clause, GLOBAL_SIGNALS would be directly visible. It is left as an exercise for the reader to discover why

use GLOBAL_SIGNALS.CLOCK;
use PROJECT_LIB.GLOBAL_SIGNALS;

would not accomplish the same thing.

In general, a use clause may take one of three forms:
- *library_name.package_name*
- *package_name.object*
- *library_name.package_name.object*

The portion of the name preceding the final period (".") is called the *prefix*; the name at the end is called the *suffix*. The suffix, which must be declared within the package named by the prefix, or contained within the library named by the prefix, is then made directly visible within the name space in which the use clause appears.

The suffix of the name in the use clause may also be the reserved word **all**. In this case, the meaning is that all declarations in the preceding package declaration, or units in the preceding library name, are made directly visible, just as if multiple use clauses, one for each declaration in the package declaration or unit in the library, had been given. Thus,

      use PROJECT_LIB.GLOBAL_SIGNALS.**all**;

would make not only CLOCK, but all other signals declared in the package declaration for the GLOBAL_SIGNALS package, directly visible.

Use clauses may generally appear wherever declarations may appear. Since multiple use clauses may appear at a given point, the possibility exists, especially with the use of the **all** option, that more than one object with the same name might be made directly visible — ironically, the problem with name space independence that we were trying to solve in the first place. In this case, where two or more objects with the same name might be made visible, the problem is solved by making neither of them visible. This is not an error, though a friendly tool might inform the user of the situation.

Objects in a design library are divided into two categories: primary and secondary units. The primary units are the package declarations, entity declarations, and configuration declarations, while the secondary units are architecture and package bodies. Note that the secondary units do not export any names; that is, all names that are made externally available come from primary units. Thus, only primary units may be named in the use clause for a particular library.

There are other visibility rules in the language, relating to some of the other name spaces; these rules are generally obvious from the rest of this book and are not explicated here. For example, when providing a parameter list, the associations may be provided by giving the names of the formal and the actual together, connected with the "=>" symbol. Clearly, the name of the formal parameter must be visible at this point. Most of these cases are simply legalisms to make the language hold together as an abstract description. The discussion above, however, is important to a clear understanding of how to write good designs in VHDL.

Access to External VHDL Libraries

When a design entity is analyzed, it may, as has been mentioned above, have access to information defined in any number of design libraries. The logical names for these libraries are given by a *library clause* immediately preceding the design entity. This library clause takes the form

> **library** *list of logical names* ;

Each named library is located by the analyzer according to the rules established by the particular implementation, and names of primary units declared in the design library are made visible by selection (*i.e.*, with a prefix-suffix notation) at all places within the design entity that the library clause precedes.

Design libraries visible within a particular entity are divided into two classes: a working library (WORK), and a number of resource libraries. The working library is the library into which the results of the analysis process (typically some kind of intermediate representation) are placed. The resource libraries contain other information, such as entity and package declarations, that is referenced by the unit under analysis. Thus, the resource libraries are not written into by the analysis process; in fact, the assumption is made that the contents of the resource libraries do not change during the analysis. This assumption is made for the working library as well, except that of course the results of the analysis are written into the working library.

Since each analysis possesses a working library, the language automatically makes the working library visible to the unit under analysis by implicitly including a

> **library** WORK;

clause immediately prior to the start of the unit declaration. Note, however, that components of library WORK are not directly visible; to accomplish that, a **use** clause for the WORK library would be required.

Since each analysis uses types and objects provided by the STD library, that library as well is made visible to the unit under analysis. In this case, a

> **library** STD; **use** STD.STANDARD.**all**;

clause is implicitly included. This makes all elements of library STD visible within the analysis, and also makes all declarations within package STANDARD directly visible (*i.e.*, without providing their full names). Obviously, STD is a resource library.

If other resource libraries are desired for a particular analysis, they must be explicitly made visible via library clauses.

Partitioning a Design

While it may be reasonable to build or document an entire design for a relatively small component in a single entity, once the size of the object being designed grows beyond a few gates, the complexity of this effort becomes difficult to manage. One of the major goals of VHDL is to support the partitioning of a design into several pieces, each of which is independently designed and described, and to then allow the convenient assembling of these piece parts into a configuration of the whole. This section discusses the two major decomposition aids in the language.

VHDL, being a language intended to support the design of complex hardware systems, naturally contains a mechanism to decompose hardware into lower level components. This mechanism, discussed in Chapter 6, is the design entity. Design entities are referenced, or instantiated, within another entity by means of the component instantiation statement; a component instantiation statement is normally said to define a subcomponent of the entity in which the statement appears.

Since complex hardware systems typically have complex algorithmic behaviors, it is reasonable to expect VHDL to also contain a mechanism to decompose algorithmic hardware behavior descriptions; this mechanism is the subprogram. Subprograms may be referenced, or called, within an entity or another subprogram by means of a subprogram call, or invocation. Subprograms have a number of uses within the VHDL language. For example, function subprograms may be used as bus resolution functions, which determine the value of a network that may be driven by multiple sources. In the context of this chapter, however, the discussion centers around the use of subprograms as decomposition mechanisms.

There are many parallels between the definition and use of subcomponents and the definition and use of subprograms. Each is primarily intended to support design decomposition for clarity, reusability, or division of labor among multiple designers. Each allows the passing of parameters of one kind or another. Component instantiations allow both generic parameters, which personalize a (sub)component for a particular use, and signal parameters (*i.e.*, ports), which specify the signals that will be connected to specific ports of a (sub)component in a particular instance. Subprogram calls allow parameters, which specify data to be passed into the subprogram to be operated upon.

However, subcomponent instantiation should be regarded primarily as a method of decomposing structure, while subprogram invocation is a method of decomposing behavior. That is, subcomponent instantiation is a relatively static process: since it reflects structure, it does not itself have execution semantics, and subcomponent parameters represent static connections. Subprogram invocation, on the other hand, carries with it the connotation of activity occurring at model execution (simulation) time. At a more language-oriented level, the two are also rather different, since a component instantiated by a component instantiation statement may update globally declared signals, such as clocks, while a procedure call may not (see the next section for more discussion of this point). Another way of looking at it is simply to note that the purpose of component instantiations and separate entity declarations is in fact to partition a design; thus, it makes sense for these declarations to be completely general and to have access to all global information. Procedure declarations, on the other hand, are intended as a way of reducing the complexity of a description, and are thus inherently more local structures; it would only be confusing to the logical model of the language if they were given access to global signal structures.

Concurrent and Sequential Procedure Calls

The procedure call statement, as with many other statements in VHDL, occurs in two forms: concurrent and sequential. A sequential procedure call statement is much like a procedure call statement in any other language. In VHDL, it may appear within a process, or within another procedure or function.

In keeping with the general philosophy of defining all concurrent statements in VHDL in terms of process statements, a concurrent procedure call is in fact equivalent to a process statement with no sensitivity list, a body consisting of a sequential procedure call to the given procedure with the given parameter list, and a terminating **wait** statement naming each signal of mode **in** or **inout** that is passed to the procedure. So, for example, the procedure call

    my_proc(in_signal, out_signal, in_variable) ;

is equivalent to the process statement

    process
    begin
        my_proc(in_signal, out_signal, in_variable);
        wait on in_signal;
    end process ;

or, given the equivalence of **wait** statements and sensitivity lists, to the process statement

    process (in_signal) begin
        my_proc(in_signal, out_signal, in_variable);
    end process ;

Thus, a concurrent procedure call statement will execute whenever any of its input signals changes, and at no other time. Note that even if other (global) signals are referenced in the procedure, changes in those signals will not result in procedure activation, since they do not appear in the sensitivity list of the process statement. In this way, the sensitivity of a procedure is determined by its parameters, without reference to its internals.

A concurrent procedure call statement (in fact, any process) is said to be *passive* if neither it, nor any procedure or function it calls, contains a signal assignment statement. Such a procedure call cannot affect its signal environment, but may well have access to signals in its environment. These procedures may be used within an entity declaration to evaluate assertion conditions that are too complex to be stated directly in the entity declaration, or to express assertion conditions that are normally parameterized, such as setup and hold times (Chapter 10, VHDL In Use, illustrates this).

Signals and procedures may be declared within packages (see below); as a result, a procedure may be invoked within a process statement without being declared itself within that process statement. This could result in an interesting situation in which the driver that was affected by a given signal assignment statement could not be determined at compile time, thus creating great problems for the simulator builders.

This could arise as follows. Remember that VHDL's model of signal values is based on the concept of drivers; a driver for a signal is created whenever a signal assignment with that signal as a target occurs either in a process statement or in a procedure invoked from that process statement. Suppose that within a package, a global bus signal were declared, along with a procedure that drives the bus by containing a signal assignment statement whose target is the bus signal. Under these circumstances, the simulator writer could not associate the code that simulates that procedure body with any particular driver; the driver is determined only for each invocation of the procedure from a given process. If the target signal were instead passed in as a parameter to the procedure, the driver is simply the driver in the process statement that invokes the procedure, and the simulator builder can pass driver information as part of the parameter.

Note that this problem does not exist if the procedure in question is defined inside a process statement, since in that case, it can be invoked only from within that process statement (it is not visible outside the process statement). The result is that the only driver for the global signal that is created is the driver associated with the process statement.

In order to solve this rather serious problem for the simulator builder, VHDL imposes restrictions on signal assignment statements in procedures. Specifically, if a procedure, P, is not declared inside a process statement, then any signal that is the target of a signal assignment statement in P must either be a formal parameter of P, or else must be a formal parameter of some other procedure *inside which P is declared.*

In summary, if a procedure makes an assignment to a signal, and if that signal is not a formal parameter of the procedure, then both the procedure and the signal must be declared within a (the) process statement that invokes the procedure.

As an aside, during the IEEE deliberations on the language standard, there was considerable debate regarding whether signal declarations themselves should be allowed in packages. One problem was the one just discussed; another, perhaps equally serious problem, is the connectivities in the networks that are thereby established without benefit of the documentation provided by ports and parameter lists. This is discussed some more in the section on Packages, below. Finally, a compromise solution was reached, in which the above restriction on assignment was included, but no restrictions were placed on signal declaration in a package or the reading of such signals; the benefit of being allowed to declare global signals such as clocks, power and ground was determined to outweigh the problems.

The Block Statement

Before discussing the component instantiation statement as a means of design decomposition, we first make a short digression to address the block statement. Like many programming languages, VHDL provides a means to set off a part of a design by labeling it and enclosing it in syntactic brackets, in this case **block** and **end block**. In the simplest case, these brackets have no effect whatsoever on the meaning or interpretation of the model in which they appear; for example:

        S <= '0' **after** 4ns;
        L1: **block begin**
            T <= '1' **after** 5ns;

            ...
        end block;

is exactly equivalent to:

        S <= '0' **after** 4ns;
        T <= '1' **after** 5ns;

However, blocks may also have a declarative part, which comes between the **block** and the **begin**. This declarative part may declare objects of various kinds, including signals. It may also define the "interface" to the block, in that it may define ports and generics of the block by providing a **port** interface list and/or a **generic** interface list.

Finally, in order to say how the external environment of the block is mapped onto those ports and generic parameters, a **port map** association list and/or a **generic map** association list may be provided. These lists, which are normally seen only in a component instantiation or a configuration specification, are placed here since there is no other convenient place to put the association between the actual signals and generic parameters and the formal names that may be used within the block.

This may all seem rather artificial; who would set up such a complicated mechanism when no benefit in terms of sharing or reuse is gained thereby? (Remember, this block is physically inside of some architecture, and may not be used outside of that architecture). Indeed, this is a valid objection. It is very rare to find a block written this way. The major reason for allowing formal ports and generics, as well as port maps and generic maps, on blocks is to provide for an equivalence between component instantiations and blocks. This is discussed in the next section.

One final feature of the **block** statement is important and worth mentioning here; this is the idea of a *guard*. A guard is a boolean expression. If a block is to have a guard, the guard must be given immediately after the reserved work **block**. Thus, for example:

> RISING_EDGE: **block** (CLOCK'EVENT **and** CLOCK='1')
> **begin**
> ...
> **end block** RISING_EDGE ;

The effect of this block guard is to define, implicitly, a signal named "GUARD" within the declarative portion of the block statement. This signal may be used as may any other signal within the block, except that it may not be driven within the block. The implicit GUARD signal, if it exists, is used to control the operation of certain signal assignment statements within the language, known as *guarded signal assignment statements*. These are discussed in detail in Chapter 9, Advanced Features; their purpose is to provide for explicit disconnection of a source from the network that it drives, or simply to gate the execution of such a signal assignment statement by the value of the boolean GUARD.

Component Instantiations and Blocks

A strong analogy can be constructed between the process of instantiating a component and selecting a chip to go into a particular socket on a board. The component instantiation statement, together with the component declaration, define the interface that the board is providing to the chip (that is, they define the socket). The configuration specification defines the actual chip that will be plugged into the socket. The various aspects of a configuration specification correspond to various levels of definition of that chip. A configuration specification takes the form:

> **for** C : CPU **use entity** mc68020
> **generic map** (*list of generic parameter associations*)
> **port map** (*list of signal/port associations*)

In this example, "C : CPU" refers to the component instantiation statement, C being the label on the statement. "mc68020" identifies the chip family being used. The generic map clause could, in this case, identify aspects of the chip such as its clock rate (by a clause such as "speed => 25 MHz", if "speed" were a generic parameter of the mc68020 model) or packaging. In other examples, generic maps could be used to specify other aspects of the chip family, such as gate timings, multiplier widths, or the order of an FFT. Now that the chip is fully defined and selected, the port map clause defines how it is wired up by specifying how the pinout of the chip maps to the wires available in the socket. A component instantiation is thus a static operation, in the sense that its definition is part of the model itself, not something that is evaluated at simulation time.

Except for issues of simplicity of design and simplicity of description, a chip placed on a board, or a macro cell designed into a chip, can be replaced by equivalent random logic at the same level as the rest of the component. In the same way, every VHDL component instantiation is equivalent to a collection of in line VHDL code, in the form of block statements, that describes the behavior of the subcomponent without using the component instantiation statement. The mapping between these two equivalent forms of description is made precise by the language.

However, the benefits gained from the decomposition into subcomponents in a VHDL description are every bit as valuable as those gained in the world of physical design — not surprising, since VHDL is intended to support precisely those kinds of design applications that require decomposition. When building hardware, macro cells or chips are often treated as "black boxes"; that is, their function is used and assumed, but their structure is mostly irrelevant. Similarly, VHDL subcomponents may be used in the construction of more complex designs without requiring the designer to fully understand, or even to know, the structure of the subcomponents.

The remainder of the discussion in this section describes the transformation of component instantiations into equivalent in-line code. It is particularly technical, and focuses directly on issues that are almost pure language issues. The reader who has grasped the essential idea of this section, that component instantiation is a "shorthand" way of writing descriptions which provides benefits in terms of simplification, reuse, and sharing, and who feels no need to understand the details of the language itself in this area, can skip the remainder of this section.

The instantiation of a component, when complete, specifies not only the entity to be used, but also a particular body of that entity. There are thus three declarative regions to be dealt with in the "flattening" of this instantiation: the declarative region surrounding the instantiation statement, the declarative region of the entity declaration itself, and the declarative region of the chosen body. There are also two statement regions to be dealt with: the statements that may appear in the entity declaration, and those that may appear in the architectural body. Finally, there is information in the component declaration within the instantiating entity, information given by a configuration specification in the instantiating entity, and information given in the component instantiation statement itself, that must be used to determine the mapping of ports and generic parameters.

To make sense of all of this, two nested blocks are introduced. The declarative portion of the first maps actual signals and generic values to the local names given in the component declaration; its body consists of the second nested block statement. This inner block statement represents the merging of the information given by the entity and architectural body declarations. Its declarative portion contains, first the mapping of the local names from the component declaration to the formal port and generic names given in the entity declaration; the remainder of the declarative region consists of the other declarative items from the entity declaration, followed by the declarative items from the body. The statement portion of this inner block contains the statement part, if any, of the entity declaration, followed by the statement part of the chosen body. Note that merging the declarative regions in the entity declaration and in the body is possible because of the scoping rules of the language — items declared in the entity are visible within each body anyway, so no conflicts in naming are possible as a result of this merging.

A simple example will perhaps make this process somewhat clearer. Suppose we had the following component declaration, component instantiation, and configuration specification within a particular architecture:

```
component adder_8 port ( i1, i2: in BV_8;
      sum: out BV_8;
      carry: out BIT );
end component;

for a8 : adder_8 use
      entity adder_cell( ripple_carry )
      port map (P1=>i1, P2=>i2,
```

                        P3=>sum, P3=>carry);

a8: adder_8 **port map** (i1=>S1, i2=>S2,
                        sum=>T, carry=>CY);

where i1, i2, sum and carry are local ports, P1, P2, P3 and P4 are formal ports, S1, S2, T, and CY are actual ports, and BV_8 is a type that is defined to be

                        array (1 **to** 8) **of BIT**;

Assume further the following entity declaration and architecture for the adder_cell:

        entity adder_cell **is**
            port(
                P1, P2: **in BV_8**;
                P3: **out BV_8**;
                P4: **out BIT**);
        end adder_cell;

        architecture ripple_carry **of** adder_cell **is**
            signal internal_carry : BIT;
        begin
            ... -- body of ripple carry adder
        end ripple_carry;

Then the transformations just described would produce the following nested block structure in place of the component instantiation labeled "a8":

        a8: **block**
            port (
                i1, i2: **in BV_8**;
                sum: **out BV_8**;
                carry: **out BIT**);
            port map (
                i1=>S1, i2=>S2,
                sum=>T, carry=>CY);
        begin
            adder_cell: **block**
                port(
                    P1, P2: **in BV_8**;
                    P3: **out BV_8**;
                    P4: **out BIT**);
                port map (
                    P1=>i1, P2=>i2,
                    P3=>sum, P3=>carry);

```
        signal internal_carry : BIT;
    begin
        ... -- body of ripple carry adder
    end block adder_cell;
end block a8;
```

As a final note, although the amount of code generated via these transformations by a simple component instantiation is staggering, remember that much of it is simply notational, and should, in a good implementation, result in no added overhead at simulation time. Also, this expanded version should never be seen by humans; it is explicated here in order that the interested reader may understand the fundamental semantics of the component instantiation statement.

Sharing Data Within a Design

As discussed previously in this book, types and objects in VHDL must be explicitly declared; once declared, a name is generally visible only from the point of declaration until the end of the declarative region within which it was declared. If data is to be used by more than one entity within a design, therefore, it must be declared in a package, since package declarations may be used to export names, and thereby objects, to the outside world (as was discussed in the visibility section above).

A VHDL package is, in fact, simply a collection of declarations. A package is divided into two parts: the "declaration" and the "body". The package declaration declares those parts of the package that are to be available for use by the "outside world", *i.e.*, by the designs that use the package. The package body provides additional declarations; for example, it provides the definition for functions whose interfaces are declared in the package declaration.

Most VHDL declarations are legal within a package declaration; the notable exception is the variable declaration, which is omitted since it leads to obvious problems with simultaneous update, provides communications paths within a design that are difficult to understand in terms of hardware, and generally results in poor design and modeling practices. Also, entities may not be defined within a package, though component declarations for externally defined entities may be given. Note also that the restrictions previously described regarding signal assignments to globally declared signals (which *may* be declared in packages) are also present partially to force improved clarity of the design description.

The remainder of this section discusses a number of possible ways in which packages might be used, and provides some examples.

A package can be used to define a number of types, and to define operations on those types. For a large project, it is likely that one or more project-level abstractions might be defined. For example, global signals or busses used throughout the project might be defined in a package, together with their bus resolution functions.

In addition to such project-level packages, technology-level packages are also useful. For example, a technology-based standard set of logic levels, together with an algebra defining operations between operands of that type, could be placed in a package for use by all members of a project or all users of a standard library. These types and operations could even be placed in a package together with component declarations for the standard components that use the types.

If the package provides all the required operations on the type, it can be thought of as abstracting the actual definition of the type away from the design itself in the sense that the users of the type need not understand how objects of the type are constructed — all that need be understood is what operations are supported on the type. This is known in the language and software business as a "data abstraction".

The example below shows a small package defining types, operations, and standard components for a four-valued (0, 1, Z, X) logic. The components are all provided with generic values for controlling various transition times — low to high, high to low, clear, and preset.

```
package Four_Valued_Logic is
  type Bit4 is ('0', '1', 'Z', 'X');
  type Bit4_Vector is array(<>) of Bit4;

  function "and" (L,R: Bit4) return Bit4;
  function "or" (L,R: Bit4) return Bit4;
  function "nand" (L,R: Bit4) return Bit4;
  function "nor" (L,R: Bit4) return Bit4;
  function "xor" (L,R: Bit4) return Bit4;
  function "not" (L: Bit4) return Bit4;

component AND_GATE
  generic (tPLH: Time; tPHL: Time);
  port (INPUT : in Bit4_vector; OUTPUT : out Bit4);

component BUF
  generic (tPLH: Time; tPHL: Time);
  port (INPUT : in Bit4; OUTPUT : out Bit4);

component DFF
  generic (tPLH,tPHL: Time; tPCL,tPPH: Time);
  port (D,CLK,PRESET,CLEAR: in Bit4;
        Q,QB: inout Bit4);
```

```
component INVERTER
  generic (tPLH: Time; tPHL: Time);
  port (INPUT : in Bit4; OUTPUT : out Bit4);

component JKFF
  generic (tPLH,tPHL: Time; tPCL,tPPH: Time);
  port (J,K,CLK,PRESET,CLEAR: in Bit4;
           Q,QB: inout Bit4);

component NANDL
  generic (tPLH: Time; tPHL: Time);
  port (SET,RESET : in Bit4; Q,NOT_Q: inout Bit4);

component NORL
  generic (tPLH: Time; tPHL: Time);
  port (SET,RESET : in Bit4; Q,NOT_Q : inout Bit4);

-- other 4-valued logic components might be defined here

end Four_Valued_Logic;


package body Four_Valued_Logic is

  function "and"(L,R: Bit4) return Bit4 is
  begin
    if L = 'Z' or R = 'Z' then return 'Z';
    elsif L = '0' or R = '0' then return '0';
    elsif L = 'X' then return 'X';
    else return R;
    end if;
  end;

  function "or"(L,R: Bit4) return Bit4 is
  begin
    if L = 'Z' or R = 'Z' then return 'Z';
    elsif L = '1' or R = '1' then return '1';
    elsif L = 'X' then return 'X';
    else return 'R';
    end if;
  end;

  function "nand"(L,R: Bit4) return Bit4 is
  begin
    return not ( L and R );
  end;
```

```
function "nor"(L,R: Bit4) return Bit4 is
begin
  return not ( L or R );
end;

function "xor"(L,R: Bit4) return Bit4 is
begin
  if L = 'Z' or R = 'Z' then return 'Z';
  elsif L = 'X' or R = 'X' then return 'X';
  elsif L = '0' then return R;
  else return not R;
  end if;
end;

function "not"(L: Bit4) return Bit4 is
begin
  case L is
    when 'Z' => return 'Z';
    when '1' => return '0';
    when '0' => return '1';
    when 'X' => return 'X';
  end case;
end;

end Four_Valued_Logic;
```

Note that in the above example, the package body defines the behavioral operations (functions) that provide the algebra for the four valued data type; however, the bodies of the entities that define the basic hardware components modeled using this data type are not defined in the package. To understand this, remember that the component declarations that appear in the package specification are merely templates for actual hardware models; these models are bound to the declarations either directly within a design entity that uses one of these components, or within a separate configuration declaration (see a later section in this chapter for a more detailed discussion of these concepts).

Another interesting point about this example is the names of the functions declared. These functions are named such things as "and" and "or", which are also names of reserved words in VHDL. This is an example of a concept called *overloading*, which means that several procedures or functions with the same name may be visible (even directly) at the same point, and the identification of which one is being referred to is made by examining the parameters and return value of the subprogram call. Overloading is discussed in more detail in Chapter 9, Advanced Features.

Although this example of a technology package does not include signals, it is perfectly legitimate for signals to appear in packages. Although, in general, placing signal definitions in packages tends to muddy (for the human reader) the question of what entities within a design use which signals, there are cases in which this may be the only reasonable alternative. For example, in a clocked design, unless the definition of the clock signal is placed in a package, it must be explicitly imported into each entity in the design via a port. This gets unwieldy for large designs. On the other hand, it should be noted that placing the definition in a package tends to decrease the reusability of each design entity, since it refers explicitly to such a global signal; thus, use of global signals should be made sparingly, and only when the benefits to model definition size and complexity clearly outweigh the costs in terms of reusability and human readability. When using global signals, however, it must be remembered that there are restrictions in the language regarding assignments to drivers of these signals; these restrictions were discussed in a previous section of this chapter.

Specifying a Design Configuration

Previous sections have discussed aspects of the language that support design decomposition in various ways: decomposition mechanisms, information sharing in a decomposed design, and management of the resulting complexity. This section addresses a different aspect of design decomposition by focusing on the issue of how the full design is reconstructed once the various parts defined by the decomposition have been fully designed. This reconstruction is necessary for various purposes, including simulation, netlist extraction, or delivery. An important aspect of VHDL is the way in which it makes it easy to provide different views of a single design by configuring it differently, without requiring re-analysis of any part of the design itself. This capability also allows rebinding of a design to incorporate a different implementation of a subcomponent — for example, a bought versus a custom-designed part.

By way of introduction, recall that when a (sub-)component is declared for use within a particular entity, what is actually declared is a template, specifying the port and generic parameter interface to the subcomponent. Further, the component declaration may be bound to a specific entity in the library by means of a configuration specification that appears within the same architecture. The component may be rebound to a new implementation by altering the configuration specification and re-analyzing the instantiating component.

This capability is useful, but it still requires alteration of the instantiating model itself, thus reducing the model's reusability and also introducing issues of configuration management that would be better avoided. The configuration declaration, mentioned briefly during the discussion of configuration specifications in Chapter 6, takes this concept one step further by allowing the designer to collect all of the bindings for a model into a single place; as a result, the particular instances of various components in the system can be changed without forcing alteration or reanalysis of any component of the system. In a way, the configuration declaration represents the high-level schematic of a system; altering the configuration declaration could be analogous to replacing one or more chips on the board with other (hopefully equivalent) chips or components.

The configuration declaration thus provides powerful binding capabilities that allow the designer to more dynamically create the relationship between a component template used in an architectural body and an actual design unit which exists in the library environment. However, if one thinks about the process of defining specific components to be used at all levels of a large model, it is clear that the process involved is a rather complex one, since choices made at a higher level (which subcomponent is selected for use) affect the choices that need to be made at lower levels: for example, selecting a purely behavioral model at a high level removes the need for further choices, while selecting a purely structural model requires that that model as well be configured.

How Component Binding Occurs

A configuration declaration begins by giving the configuration a name, and associating it with a particular entity; thus, each configuration declaration defines a configuration for a particular entity. For example, a configuration declaration beginning

configuration FULL_SLOT **of** COMM_BOARD **is**

could go on to define a configuration for COMM_BOARD that fits into a full PC/AT slot.

Following the initial line of the declaration can be a number of use clauses to gain visibility into packages for types, data items, or functions as needed. Attribute specifications may be given in the component declaration as well, and may be used to distinguish particular instances of a component. For example, if the entity declaration for COMM_BOARD contained an attribute declaration named COMPONENT_COUNT, the FULL_SLOT configuration might contain an attribute specification such as

attribute COMPONENT_COUNT **of**
COMM_BOARD : **entity is** 250;

This information could be used by management tools to compute total part count.

After these initial declarations, the real work of the configuration declaration is done by a number of additional statements, each of which provides a configuration for one or more of the blocks or components in the entity. Remember, however, that the entity being configured will, after configuration is complete, be completely elucidated, i.e., no unbound component declarations will remain. Thus, even if the base entity (COMM_BOARD in our example) contains instantiations that themselves contain subcomponents, the configuration declaration must provide a way to descend that tree, through all of its levels. Thus, in our example, we might say

```
        configuration FULL_SLOT of COMM_BOARD is
          attribute BOARD_KIND of COMM_BOARD : entity
                               is FULL_SIZE;

        for LOW_COST -- an architecture of COMM_BOARD

            for CPU : PROCESSOR
              use entity STD_PARTS.SPARC( Fujitsu )
                generic map (Clock => 40 ns);
            end for; -- for PROCESSOR

            for BUS_CONTROLLER
              for Protocol_Chip : IEEE_488
                use entity STD_PARTS.NEC ;
              end for;

            end for; -- for BUS_CONTROLLER


            ...
        end for; -- for LOW_COST
        end FULL_SLOT;
```

In this (almost) complete example, there is a single block configuration (starting **for** LOW_COST) within the configuration declaration; within that block configuration are nested two other *configuration items*. The outer block configuration states that the LOW_COST architecture of COMM_BOARD is the one configured by this configuration declaration. If a block configuration identifying an architecture does not appear, the language provides default rules to select an architecture. These default rules are discussed further below.

There will likely be many subcomponents on our example board. Two have been singled out here: the PROCESSOR and the BUS_CONTROLLER. The board in this example is an IEEE 488 controller board with an on-board processor for doing instrumentation data acquisition and processing. The first configuration item inside the block configuration is a *component configuration*, which is used to select the processor, which in this case is the SPARC machine. We choose the Fujitsu implementation (which is signified by an architecture for SPARC whose name is "Fujitsu"), and genericize the binding indication to indicate use of a 25 MHz part (a 40 nanosecond clock rate). There must be, of course, a component instantiation statement for PROCESSOR in the LOW_COST architecture for COMM_BOARD. The label of that statement is "CPU". Note that one would not normally expect the identity of the processor itself to be selectable at such a late stage in the design; however, it is not unreasonable to suppose that the manufacturer, and certainly the clock rate, are so selected.

The second configuration item inside the outermost block configuration is another block configuration, which is used to provide a configuration for a nested block in the architectural description of COMM_BOARD. In our example, we are here assuming that the bus controller is described inside the LOW_COST architecture as a bunch of random logic together with an IEEE 488 controller chip, and that all of this description is contained in a block statement named "BUS_CONTROLLER". The only purpose of this block configuration is to open up the name scope of the block so that the component instantiation for the IEEE_488 controller chip is made visible. That chip is configured using the NEC IEEE 488 part, through the use of a nested component configuration.

Just as the block configuration may have block or component configurations nested within it, so may the component configuration have a block configuration nested within it. In fact, the only way to open the scope of the architecture configured with a component configuration is to nest a block configuration naming the architecture within the component configuration. Thus, if it had been desired to further configure components (or blocks) nested within the Fujitsu implementation of STD_PARTS.SPARC, the designer would have written:

```
        for CPU : PROCESSOR
            use entity STD_PARTS.SPARC( Fujitsu )
                generic map (Clock => 40 ns);

            for Fujitsu
                -- component configurations for instantiations in Fujitsu, or
                   block configurations for blocks nested in Fujitsu
            end for;
        end for; -- for PROCESSOR
```

There is an obvious similarity between a component configuration and a configuration specification appearing in an architecture. They differ only in that the component configuration allows the nested block configuration, and is terminated with an **end for**. This difference between the component configuration and the configuration specification is precisely what allows the component configuration to provide access for configuration "down the hierarchy", and is thus the root of the additional capabilities of the configuration declaration over configuration specifications in architectures.

An interesting feature of the COMM_BOARD example that may not be obvious at first glance is that, since the name "PROCESSOR" is defined internally to the architecture, one would not normally expect it to be visible to the outside world. However, the "for LOW_COST" clause has opened up the scope of the architecture, making names defined within it visible from within the "for LOW_COST" statement. This is how the language solves the problem of how to allow configuration arbitrarily far down a design tree when normally design decisions made within a particular architecture are regarded as private and thus not visible from without the architecture.

If, in our example, BUS_CONTROLLER had actually been written as a component instantiation that referred to a region of the board (remember that a component instantiation is equivalent, in many ways, to a nested block statement), then we would have first had to bind BUS_CONTROLLER to a particular implementation for this function (which could be thought of as a macro cell at the board level), and then bind the IEEE_488 subcomponent within it. This could have been written:

```
for B1 : BUS_CONTROLLER
    use entity
    COMPANY_LIBRARY.STD_CONTROLLER( Std_body );
    for Std_body
        for Protocol_Chip : IEEE_488
                use entity STD_PARTS.NEC ;
        end for;
    end for;
end for;
```

Note the similarity between the two configurations; the difference (the additional use statement and the nested block configuration) is explained by noticing that in the first case, the identity of the block was known without further discussion; in the second case, since a component instantiation is used, the name of the actual library component to bind to must be given.

It should be noted that the "use binding-indication;" phrase in a component configuration inside a configuration declaration may (and in fact must) be omitted if there is a configuration specification for the component in the architecture being configured. In this case, the architecture inside the component must be opened as usual with a block configuration; however, the identity of the component bound to is established through the configuration specification.

A final capability of configuration declarations that is well worth mentioning is the ability to configure a part with a configuration declaration and then to use that configuration declaration as the binding indication for a component instantiation naming its entity. For example, a complete system that made use of COMM_BOARD might, in many situations, wish to be unconcerned with the configuration details of the board, but would know only that the board must fit in a half-slot (PC/XT slot). In this case, the user of the board could simply write, in a configuration specification (which could appear within a configuration declaration or within an architecture):

> **for** COMM_BOARD
>   **use configuration** HALF_SLOT;
> **end for**;

Once the tree of configuration specifications within the configuration declaration has been fully walked, and configuration decisions made, it is still possible that component instantiations will remain unbound because no configuration specification was provided. In this case, the configuration binding is controlled by default rules specified in the language. These rules state that, for each unbound component instantiation, an entity whose name is the same as the name given in the component declaration inside the architecture is located in the library (if one cannot be found, an error occurs). The most recent architecture of that entity is identified, and that is used to bind the component instance. The port and generic maps are matched by name.

This defaulting facility provides a clean way of incorporating standard parts libraries without the need for explicit specification within a configuration declarations, and by the use of standard names. A (largely empty) configuration declaration is, however, required in order to get the defaulting process going. For instance, if, in our example above, there was a standard IEEE 488 chip in use in a particular organization, the configuration specification for it would not have been needed; the default rules would have located the chip, named IEEE_488, and used it in the binding.

The rules of configuration declarations are complicated, and provide a lot of capability to the user. However, this capability is extremely useful, and the flexibility gained is immense. The judicious application of the defaulting rules in the language can greatly reduce the amount of configuration data that must be created, but the full generality is always present if needed.

Type Incompatibilities in Component Binding

The previous section discussed the use of configuration declarations to resolve name binding issues in binding component instantiations. The configuration specification can also be used, as was discussed in Chapter 6, to resolve incompatibilities such as unconnected ports, port names, or port ordering.

The purpose of this section is to discuss another incompatibility that may be resolved through the use of the configuration specification: incompatibilities in the types of the ports. This can become a real issue in the case where some set of models has been acquired from outside the organization developing a piece of hardware, and it is desired to use those models. In this case, the developers of the model library undoubtedly had their own ideas about what data types allowed them to most accurately model the behavior of the components in their library; these data types will likely not correspond to the set of data types used by the hardware designers. If the parts are to be usable in the VHDL models without a lot of work (work that may not be able to be performed by the hardware designer, since his organization may not have access to source code for the models), VHDL must provide some way to make these two sets of types "talk" to one another. VHDL functions are used to solve these problems; when a function is used in this manner, it is commonly called a *type conversion function* or a *type transformation function*.

For example, suppose one was to purchase a set of models written in VHDL that used the Bit4 data type discussed previously, with the algebra as described in the Four_Valued_Logic package. That data type is appropriate for modeling hardware at the level at which one is concerned about undefined or in-transition values; users of the components are probably modeling at the functional level, where only BIT (binary) values are important or interesting. Thus, there is need in this example for a transformation between Bit4 values and BIT values. Two VHDL functions can easily be written to describe these transformations:

```
function Bit4_to_BIT( b : in Bit4 ) return BIT
is begin
  case b is
    when '0' => return '0';
    when '1' => return '1';
    when 'X' => return '0';
    when 'Z' => return '0';
  end case;
end Bit4_to_BIT;

function BIT_to_Bit4( b : in BIT ) return Bit4
```

```
is begin
  case b is
    when '0' => return '0';
    when '1' => return '1';
  end case;
end BIT_to_Bit4;
```

Note that some arbitrary decisions were made in the translation of Bit4 values to BIT values, since a Bit4 object has more possible values than does a BIT object. It is also interesting to note (and this topic is discussed further in the Overloading section in Chapter 9) that the literals '0' and '1', which are used in both data types, appear without any indication of the type meant by each appearance. The rules of the language are sufficient to determine the data type of each appearance of the literals.

The above example shows how different logic-level models may be transformed one to the other. Type transformation functions may be applied to much more complex situations as well. For example, one could define transformation functions to translate between 32-bit BIT_VECTORS and integers; the transformation could be defined according to the underlying arithmetic of the machine being modeled. As another example, an incoming computer instruction could be disassembled into its component parts (in an object of a record type) by a type conversion function; at the computer modeling level, the instructions are thought of as strings of bytes, while at the microcode level, they are thought of as bit fields; a particular bit field might not even consist of a contiguous set of bits within the instruction.

Now that the type conversion functions are defined, there needs to be some way to associate them with particular ports in a component instantiation. This is done (either in the configuration specification, the component declaration, or the component instantiation) by using the type conversion function names as part of the port association. Thus, for example, a signal C of type BIT that was being connected to an **in** port CLEAR of one of the components declared in Four_Valued_Logic might be associated with its formal port by the association element

$$CLEAR => BIT\_to\_Bit4(\ C\ )$$

while a signal FF_OUT of type BIT connected to the **inout** port Q of a flip-flop might be associated via the phrase

$$Bit4\_to\_BIT(\ Q\ ) => BIT\_to\_Bit4(\ FF\_OUT\ )$$

In each case, a function call syntax surrounding a name provides the conversion from the type of that name to the "other" type; each association element may have a type conversion function on either, both, or neither side of an association element.

This, then, is the means provided by VHDL to address the problem of selection of different types by different model builders. The complexity of building conversion functions is not too great, at least in cases that are likely to arise. The major issue is one of data fidelity. Whenever a conversion is made between two types with a different number of elements, there will be cases in which information is lost. This is inevitable in the general case; it is up to the designer of the conversions to ensure that the best choices are made in these cases, and that users of the models are aware of what is going on.

Mixing Structure and Behavior

Previous chapters have discussed ways to write VHDL that describes hardware from a purely structural view, and ways to write purely behavioral VHDL. The structural method often does, but need not, correspond to a design methodology in which larger parts (e.g. boards) are built up out of smaller, already defined parts (e.g. chips). Behavioral VHDL can itself take two general forms: the algorithmic, sequential form found inside a process statement, and the parallel, more register-transfer like form found directly within architecture bodies.

The VHDL language is designed to support an evolutionary design methodology, in which a single design can contain, even within a single design entity, aspects that are structural in nature and others that are behavioral in nature. This kind of description tends to evolve naturally through the process of hierarchical design of a part: as the layers of the part are successively designed, more and more of the description may be committed to behavioral description; the remaining parts are given in the design as component instantiation statements that are designed later. Finally, the component instantiation statements refer to components that exist in the library, and the design is complete. The VHDL library system, described in the previous section, allows the maintenance of all of these different versions of a design entity, and the configuration body allows the selection of different component versions for simulation.

The Traffic Light Controller example, which forms the subject matter of the next chapter, shows how this process proceeds from the definition of the entity interface to a component through a number of levels of behavioral design, to structural design, and finally to a simulatable configuration.

Chapter 8
A Complete Example

This chapter presents a single comprehensive example which illustrates the use of VHDL over most of the design cycle of a circuit. The example illustrates the development of a model from a specification into its actual implementation. A specification of a circuit is defined by an architectural body (or collection of bodies) which defines the behavior of the circuit, an entity definition which defines the interface (I/O) of the circuit, and a test bench (entity and architecture) which stimulates the circuit and captures the results in signals. In this example, a single architectural body is created which defines the behavior of the circuit with a corresponding entity and test architecture. The circuit is then partitioned and one of the partitions is broken down into a PLA circuit. The other partitions are left in behavioral form.

A nice way of putting this example in context is to think of its use for a contracted device. The contracting agent creates a specification of the system which defines the behavior of the device and a set of criteria for checking the operation of the device. This specification is then passed on to the contractor. The contractor has a simulatable description of the device to be developed. Further refinements of the specification lead to a partitioning of the device into two separate components which are to be developed by two different design teams. The interface to each component is defined and the interaction between them is placed in the behavioral description of each component. The new description is passed on to each design team. Each team then refines the description into an implementation. The beauty lies in the fact that at each of these stages there is a simulatable VHDL description of the device facilitating easy integration of the resulting components of the device. Throughout the chapter, this scenario will be re-examined in the context of the design.

The example presented here is a system which controls a traffic light at the intersection of a highway and a farm road. First, a specification is developed in VHDL which correctly models the behavior of the system. This specification is a purely behavioral model of the system instantiated by a test program written in VHDL. The specification is then partitioned and the control side of the partition is successively refined into a PLA circuit representation. This example shows how to use VHDL to perform top-down design of a circuit.

The Traffic Light Controller

The traffic light controller problem is an example of a finite state machine. The problem involves the control of the traffic lights at an intersection of a highway and a farm road. There are detectors on both sides of the farm road to detect the presence of cars. The highway light should cycle to red only when a car is detected on the farm road. The farm road light should remain green as long as cars are on the road but no longer than a given time delay. When the highway lights turn green again, they should remain green for at least a given time delay.

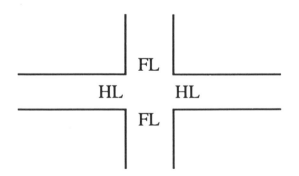

Figure 8.1. Traffic Lights at Intersection of Highway and Farm Road

The approach to designing this example will proceed in the following way:
1.   The types, subtypes and conversion functions which might be needed by various derivations of the circuit will be packaged.

2. The interface of the circuit will be defined.

3. A behavioral description of the internal activity of the circuit will be developed.

4. A test description will be developed to stimulate the circuit and record the response during simulation. The test description will be bound to the circuit.

5. The timer component of the circuit will be partitioned out of the main description into a separate design unit. A new description of the circuit will be developed which instantiates the timer component.

6. The controller component of the circuit will be partitioned out of the main circuit into a separate design unit. A new top level description will be developed which instantiates both the timer and the new controller component.

7. The implementation of the controller circuit will be begun by creating a behavioral description of the controller logic which defines the activity in terms of its bit level operations.

8. A behavioral description of a PLA circuit will be developed which expresses the function of the controller as a sum of products.

9. A behavioral partition of the PLA circuit will be developed to facilitate further structural partitioning of the PLA.

10. Finally, the PLA will be partitioned into its component parts (registers, and plane, and or plane). The further development of these units is left to the reader.

This example is meant to represent a possible approach to designing with VHDL. Other approaches are possible, for it is not the intent of VHDL to enforce a particular design methodology on to designers but to support the methodology used by the designer.

Creating the Specification

The first step in this design, as in any design, is to map out a strategy and framework for the system. A few design decisions need to be made based on the nature of the problem. Once these decisions are made the specification will follow naturally.

Defining the System Types

The first design decision is to decide the nature of the problem and, by extension, the nature of the solution. As stated above, the traffic light controller is an example of a finite state machine. This implies that a state machine representation would best suit the needs of the problem. In order to define the state machine, the states the system can be in need to be defined. The stable state of the system is when the highway lights are green and the farm road lights are red. The next state is when the highway lights turn yellow. After this state the highway lights turn red and the farm road lights turn green. In the last state the farm road lights turn yellow. This state is followed by a return to the initial state. Therefore, there are four states of the system which determine the color of each set of lights. In order to codify this information, a package is created which defines a type for the colors of the lights and a type for the state of the system.

```
-- Package for encapsulating types and constants
-- for the Traffic Light Controller Design
package Traffic_Package is

    type Color is (Green, Yellow, Red, Unknown);

    -- Type and Subtype Declarations
    type State is (
      Highway_Light_Green,
      Highway_Light_Yellow,
      Farmroad_Light_Green,
      Farmroad_Light_Yellow);

end Traffic_Package;
```

This package contains two type declarations. The first declares a type named Color which is an enumeration of all the colors that a set of lights may be. The second type defines each state the system can be in. It should be noted that when the highway lights are green the farm road lights will be red and vice versa so a separate state for a red light is not needed. equivalent.

Creating the Interface

The next design decision which needs to be made is defining the inputs and outputs of the system. The system is controlling two sets of lights and, therefore, two output signals are required: one to control the highway lights and one to control the farm road lights. The specification will contain as few design decisions as possible, so the two signals will be represented only by the color the light should be at any given point. These two outputs will be called Highway_Light and Farmroad_Light, respectively.

The system reacts (moves from its steady state) when a car is detected on the farm road. This implies that the system needs only a single input. This input will be named Car_On_Farmroad and will be of type BOOLEAN. The input will be TRUE while a car is detected on the farm road and FALSE otherwise.

There are also two constants in the system which will be set by an outside source: 1) the length of time the highway light must stay green or the maximum time the farm road light may remain green before the system moves to the next state (light goes yellow) and 2) the length of time either of the lights should remain yellow. VHDL generics will be used to allow an outside source to pass in the values. The generics will be given the names Long_Time, for the length of time a light must stay green, and Short_Time, for the length of time either light should remain yellow.

Given this definition of the inputs and outputs of the system, the following interface is defined:

```
-- entity declaration defining the inputs and outputs
-- for the traffic light controller example
use work.traffic_package.all;
entity traffic_light_controller is
  generic (
    Long_Time : Time; -- Minimum green light duration
    Short_Time : Time -- yellow light duration
  );
  port (
    Car_On_Farmroad : in Boolean;
    Highway_Light   : out Color;
    Farmroad_Light  : out Color
  );
end traffic_light_controller;
```

The entity declaration uses the data type package so that the type Color will be visible. The two generics and three ports represent the inputs and outputs of the system defined above.

The Body of the Specification

The next step is to define the behavior of the state machine. This behavior can be expressed as the activity within the system that causes the transition from one state to another.

The first state of the system is Highway_Light_Is_Green. Two aspects of the problem affect the transition of the system from this state to the state Highway_Light_Is_Yellow: the detection of a car on the farm road and the length of time the light has been green. We can state the relationship as follows: if the highway light has been green for at least the time given by Long_Time and a car is detected on the farm road, then the the highway light will move to yellow.

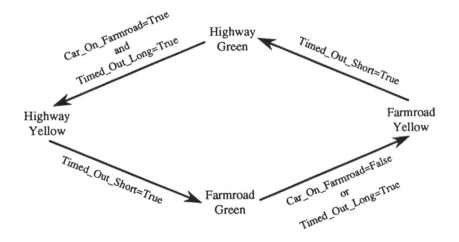

Figure 8.2. State Machine for the Traffic Light Controller

The transition from the state Highway_Light_Is_Yellow is determined only by the amount of time the highway light has been yellow. That is, when the highway light turns yellow the system will stay in the state Highway_Light_Is_Yellow until the time given by Short_Time has elapsed. Then the system will move to the state Farmroad_Light_Is_Green.

When the system is in the state Farmroad_Light_Is_Green, the transition is again affected by the detection of cars on the farm road and the duration of Long_Time. This transition is a little different than the one from Highway_Light_Is_Green and can be stated as follows: if there are no cars on the farm road or the system has been in the state for at least the time given by Long_Time, then the farm road light will move to yellow. The difference here is that in this state, when the time has elapsed the presence of cars on the farm road is ignored.

The transition from the state Farmroad_Light_Is_Yellow is identical to that from Highway_Light_Is_Yellow; namely, the transition occurs when the system has been in the state for the time given by Short_Time.

In order to determine how much time has passed once a new state is entered, the system must contain some timing indicators. Every time a new state is entered a timer will be started. The state machine will be notified when the Long_Time has passed and when the Short_Time has passed after the timer has been started. Therefore we will need three signals to provide the input and output to the timer. These signals will be named Start_Timer, Timed_Out_Long, and Timed_Out_Short to indicate their function in the timer mechanism.

The diagram in Figure 8.2 shows the state machine for the traffic light controller.

To facilitate the development of the specification, a transition table is developed. The following table is derived using the names defined above and the statement of the problem. The table shows only those transitions which change the state. The transition table is shown in Figure 8.3.

Present State	Inputs	Next State	Outputs	
			Highway Light	Farmroad Light
Highway Light Is Green	Car_On_Farmroad = TRUE and Timed_Out_Long = TRUE	Highway Light Is Yellow	Green	Red
Highway Light Is Yellow	Timed_Out_Short = TRUE	Farmroad Light Is Green	Yellow	Red
Farmroad Light Is Green	Car_On_Farmroad = FALSE or Timed_Out_Long = TRUE	Farmroad Light Is Yellow	Red	Green
Farmroad Light Is Yellow	Timed_Out_Short = TRUE	Highway Light Is Green	Red	Yellow

Figure 8.3. Transition Table for Traffic Light Controller

Using this transition table and the verbal description given above it is possible to generate a behavioral description of the traffic light controller. The specification of the traffic light controller is defined by the following architectural body.

```
-- architectural body for the specification of
-- the traffic light controller
architecture specification of traffic_light_controller is

    -- signal for holding the current state of the system
    signal Present_State : State := Highway_Light_Green;

    -- signals for implementing the timing mechanism
```

```
signal Timed_Out_Long : Boolean := FALSE;
signal Timed_Out_Short: Boolean := FALSE;
signal Start_Timer : Boolean := FALSE;

begin -- specification

  -- process statement which implements the state machine
  Controller_Process:
  process
  begin
    case Present_State is

      when Highway_Light_Green =>
        if Car_On_Farmroad and Timed_Out_Long then
          Start_Timer <= not Start_Timer;
          Present_State <= Highway_Light_Yellow;
        end if;

      when Highway_Light_Yellow =>
        if Timed_Out_Short then
          Start_Timer <= not Start_Timer;
          Present_State <= Farmroad_Light_Green;
        end if;

      when Farmroad_Light_Green =>
        if not Car_On_Farmroad or Timed_Out_Long then
          Start_Timer <= not Start_Timer;
          Present_State <= Farmroad_Light_Yellow;
        end if;

      when Farmroad_Light_Yellow =>
        if Timed_Out_Short then
          Start_Timer <= not Start_Timer;
          Present_State <= Highway_Light_Green;
        end if;

    end case;
    wait on
      Car_On_Farmroad, Timed_Out_Long, Timed_Out_Short;
  end process;

  -- Conditional signal assignment to set highway light
  Highway_Light_Set:
    with Present_State select
      Highway_Light <=
        Green when Highway_Light_Green,
        Yellow when Highway_Light_Yellow,
```

```
        Red when
          Farmroad_Light_Green | Farmroad_Light_Yellow;

    -- Conditional signal assignment to set farm road light
    Farmroad_Light_Set:
      with Present_State select
        Farmroad_Light <=
          Green when Farmroad_Light_Green,
          Yellow when Farmroad_Light_Yellow,
          Red when
            Highway_Light_Green | Highway_Light_Yellow;

    -- process statement to implement timing mechanism
    Timer_Process:
      process
      begin
        Timed_Out_Long <= FALSE, TRUE after Long_Time;
        Timed_Out_Short <= FALSE, TRUE after Short_Time;
        wait on Start_Timer;
      end process;

    end specification;
```

The architecture defines four signals. The first signal, Present_State, is used to hold the current state of the controller. It is initialized to Highway_Light_Is_Green. The next three signals are declared with a single signal declaration and define the signals passed to and from the timer process.

The architecture has four concurrent operations. The first is the process statement defining the controller element of the circuit as a state machine. The behavior of this process is dependent on the state of the controller. The case statement and the statements within each branch of the case statement directly implement the transition table. The process suspends at the bottom and waits for events on either the Car_On_Farmroad detect signal, the Timed_Out_Long signal, or the Timed_Out_Short signal. The next two statements are conditional signal assignments. They are used to convert the current state of the system into the color for the signals used to control the traffic lights themselves. The final process is the timer element of the controller. It turns the timeout signals off and then sets them to activate after the generic time delays.

Creating a Test Bench

Now that the body (architecture) of the specification has been created, a test bench should be defined to apply test inputs to the circuit. The results of these tests should be used to check the design. Once the design has been checked, the simulation results can be used as a reference to check the implementation as it is developed.

The test bench of a circuit should include signal receptors for the inputs and outputs of the circuit. It should also include test stimulation for the circuit. This stimulation is applied to the signals used for the inputs to the circuit. In many cases, it is desirable to partition the test stimulator into a separate component which is instantiated from within the test bench. This allows the designer to isolate out the specifics of deriving the test set. In this circuit however, the test stimulation can be very straightforward and therefore is included directly in the test bench.

The following entity and architecture define the test bench for the traffic light controller.

```
-- entity declaration for test bench : no inputs or outputs
entity TLC_Test is end TLC_Test;

-- architectural body for test bench for the
-- traffic light controller
use work.Traffic_Package.all;
architecture Test of TLC_Test is

  -- signal for stimulating traffic light controller
  signal Car_On_Farm_Road : Boolean := FALSE;

  -- signals for catching output of traffic light controller
  signal Highway : Color := Green;
  signal Farmroad : Color := Red;

  -- component template for instantiating a
  -- traffic light controller
  component TLC
    generic (
      Long_Time : Time;
      Short_Time : Time);
    port (
      Car_On_Farmroad : in Boolean;
      Highway_Light : out Color;
      Farmroad_Light: out Color);
  end component;
```

```
begin

    -- instantiate the traffic light controller
    Controller : TLC
      generic map (5 ns, 2 ns)
      port map (Car_On_Farm_Road, Highway, Farmroad);

    -- Stimulate the traffic light controller input
    Car_On_Farm_Road <=
      FALSE,
      TRUE  after 1 ns,
      FALSE after 3 ns,
      TRUE  after 10 ns,
      FALSE after 20 ns;

  end Test;
```

The test bench contains a signal for the detection of cars on the farm road (Car_On_Farm_Road). This is the input to the circuit and is the signal which will be stimulated by the test bench. The test bench also contains two other signals for catching the outputs of the controller which control the traffic lights (Highway_Light and Traffic_Light). These signals contain the state of the respective lights in enumeration form (Green, Yellow, or Red).

The test bench defines a component declaration for the controller circuit. The component declaration can be thought of as a template for the type of design unit that will be bound to this architecture. In this case, the component contains the same number of ports and generics as the entity of the traffic light controller and, in fact, the same names are used. It should be remembered that this is not a requirement. Differing numbers and names of ports can be resolved in the configuration of the model using the binding indication. The component is instantiated and the instantiation gives values for the generics and maps the signals to be used for input and output to the ports of the component.

Finally, test stimulation is applied to the inputs of the controller component. The signal Car_On_Farm_Road is the only input signal and only a single concurrent assignment statement is used to stimulate it.

Finally, we bind the model together with a configuration:

```
use work.all;
configuration spec of TLC_TEST is
  for Test
    for Controller : TLC use
      entity work.Traffic_Light_Controller(Specification);
```

```
        end for;
      end for;
    end spec;
```

The configuration binds the TLC component to the design unit
Traffic_Light_Controller(Specification).

The model can now be simulated. Figure 8.4 shows the expected
output from the simulation.

(NS)	Car_On_Farm_Road	Highway	Farmroad
0	FALSE	GREEN	RED
1	TRUE		
3	FALSE		
10	TRUE	YELLOW	
12		RED	GREEN
17		YELLOW	
19		GREEN	RED
20	FALSE		

Figure 8.4. Simulation Results of the Traffic Light Controller

These results can be used throughout the refinement of the design
of the traffic light controller. If the simulation results of a new
refinement do not match these results then there is an error in the
refinement.

The VHDL model now represents a complete specification of the
desired circuit. Although some design decisions have been made, no
real implementation detail has been decided upon.

Partitioning the Design

Recall from the beginning of this chapter the scenario for this
design. The contractor receives the specification defined above and is
now ready to partition the design for development by two separate teams
of engineers. One team will develop the timer section of the device and
the other will develop the controller section. In order to coordinate these

developments, the specification is partitioned into two different components. In order to check the validity of a partition, one component is partitioned out at a time and then the resulting model is simulated.

The body of the specification for the traffic light controller involves two main functions: the controller itself and the timer element. As a first pass, the timer unit will be separated into a new design unit which is instantiated from the main unit.

Choosing a Type Representation

Before partitioning out each component, a design decision about the types of data which each team will be using is made. The project manager decides that both components will be developed using the predefined type BIT. This decision has two immediate ramifications. First, the operations on the internal signals must be rewritten to operate on the BIT type. Second, the values passed from the interface to the circuit must be converted into the type BIT and the values passed back through the interface must be converted from the type BIT.

A package is created to encapsulate the definitions and operations associated with the choice of a type representation for the design. There are four states of the system and three different colors for the traffic lights, therefore, signals representing the state of the system or the color of the lights must be larger than one BIT. To handle the conversion of the inputs and outputs of the specification, a number of conversion functions are necessary. The following package encapsulates the necessary definitions and operations.

```
-- package specification for the type representation
-- chosen for the traffic light controller
use work.Traffic_Package.all;
package Design_Package is

    -- Type and Subtype Declarations
    subtype Data_Type is Bit;
    type State_Bits is array (0 to 1) of Data_Type;
    type Color_Bits is array (0 to 1) of Data_Type;

    -- Constants
    constant Green_Light  : Color_Bits := B"00";
    constant Yellow_Light : Color_Bits := B"01";
    constant Red_Light    : Color_Bits := B"10";
    constant Unknown_Light: Color_Bits := B"11";

    constant HG_State : State_Bits := B"00";
    constant HY_State : State_Bits := B"01";
```

```
constant FG_State : State_Bits := B"11";
constant FY_State : State_Bits := B"10";

-- Type Conversion Functions
function Bits_to_Color (
   Bits_In : in Color_Bits)
return Color;

function Color_to_Bits (
   Color_In: in Color)
return Color_Bits;

function Bits_to_State (
   Bits_In : in State_Bits)
return State;

function State_to_Bits (
   State_In: in State)
return State_Bits;

function Bit_To_Boolean (
   Bit_In : in Bit)
return Boolean;

function Boolean_To_Bit (
   Boolean_In : in Boolean)
return Bit;

end Design_Package;
```

The package contains two type declarations and a subtype declaration. The subtype declaration declares Data_Type to be a subtype of BIT which includes all values of BIT. The purpose of this subtype is to provide a name to be used for each declaration of an object of this type. If every declaration used the word BIT directly it would be harder to go back and change the decision of the representation type. Following the type and subtype declarations, eight constants are declared which relate the enumerations used in the specification to their bit encodings in the development. Finally, the package declares four conversion functions to convert from the bit representations to the enumeration types.

This package provides a convenient mechanism for defining the environment under which each component of the device will be developed. Without a common representation, there would be no guarantee that the work produced by each team would be in concordance with the work produced by the other team. Packaging is a very useful and important feature of the language.

The package above defined the specifications of the conversion functions. The bodies of the functions will be defined in the package body. Design units which make use of the conversion functions do not need to know about the actual implementation of each function. By isolating the functions into the package body, the implementation (functionality) of the functions can be changed without making the units which use them obsolete.

The conversion functions in the package body are straightforward and well-defined and, therefore, can be developed before the rest of the model. The following package body defines the behavior of the conversion functions.

```
package body Design_Package is

function Bits_to_Color (
  Bits_In : in Color_Bits)
return Color is
begin
  case Bits_In is
    when Green_Light =>
      return Green;
    when Yellow_Light =>
      return Yellow;
    when Red_Light  =>
      return Red;
    when others =>
      return Unknown;
  end case;
end Bits_to_Color;

function Color_to_Bits (
  Color_In : in Color)
return Color_Bits is
begin
  case Color_In is
    when Green  =>
      return Green_Light;
    when Yellow =>
      return BYellow_Light;
    when Red    =>
      return BRed_Light;
    when others =>
      return BUnknown_Light;
```

```
    end case;
end Color_to_Bits;

function Bits_to_State (
  Bits_In : in State_Bits)
return State is
begin
  case Bits_In is
    when HG_State =>
      return Highway_Light_Green;
    when HY_State =>
      return Highway_Light_Yellow;
    when FG_State =>
      return Farmroad_Light_Green;
    when FY_State =>
      return Farmroad_Light_Yellow;
  end case;
end Bits_to_State;

function State_to_Bits (
  State_In : in State)
return State_Bits is
begin
  case State_In is
    when Highway_Light_Green =>
      return HG_State;
    when Highway_Light_Yellow =>
      return BHY_State;
    when Farmroad_Light_Green =>
      return FG_State;
    when Farmroad_Light_Yellow =>
      return FY_State;
  end case;
end State_to_Bits;

function Bit_To_Boolean (
  Bit_In : in Bit)
return Boolean is
begin
  if Bit_In = '0' then return FALSE;
  else return TRUE;
  end if;
end Bit_To_Boolean;

function Boolean_To_Bit (
  Boolean_In : in Boolean)
return Bit is
```

```
    begin
      if Boolean_In = FALSE then return '0';
      else return '1';
      end if;
    end Bit_To_Boolean;

  end Design_Package;
```

Revising the Specification

Now that a type representation has been chosen, the circuit can be rewritten for the new type. This means rewriting the body of the specification to conform to the type representation and to convert values where necessary to maintain congruence with the interface of the specification. At this level, only those changes which affect the interface of the new partitions will be made; other conversions will be added as the design is developed. The new architecture for the specification is given below.

```
    use work.Design_Package.all;
    architecture revised of traffic_light_controller is
      signal Present_State : State := Highway_Light_Green;
      signal Timed_Out_Long, Timed_Out_Short : Data_Type := '0';
      signal Start_Timer : Data_Type := '0';
    begin

    Controller_Process:
    process
      variable Farmroad_Car_Detect : Data_Type;
    begin
      case Present_State is

        when Highway_Light_Green =>
          if Car_On_Farmroad and Timed_Out_Long = '1' then
            Start_Timer <= transport not Start_Timer;
            Present_State <= transport Highway_Light_Yellow;
          end if;

        when Highway_Light_Yellow =>
          if Timed_Out_Short = '1' then
            Start_Timer <= transport not Start_Timer;
            Present_State <= transport Farmroad_Light_Green;
          end if;
```

```
when Farmroad_Light_Green =>
  if not Car_On_Farmroad or Timed_Out_Long = '1' then
    Start_Timer <= transport not Start_Timer;
    Present_State <= transport Farmroad_Light_Yellow;
  end if;

when Farmroad_Light_Yellow =>
  if Timed_Out_Short = '1' then
    Start_Timer <= transport not Start_Timer;
    Present_State <= transport Highway_Light_Green;
  end if;

end case;
wait on
  Car_On_Farmroad, Timed_Out_Long, Timed_Out_Short;
end process;

Highway_Light_Set:
  with Present_State select
    Highway_Light <=
      Green when Highway_Light_Green,
      Yellow when Highway_Light_Yellow,
      Red when
        Farmroad_Light_Green | Farmroad_Light_Yellow;

Farmroad_Light_Set:
  with Present_State select
    Farmroad_Light <=
      Green when Farmroad_Light_Green,
      Yellow when Farmroad_Light_Yellow,
      Red when
        Highway_Light_Green | Highway_Light_Yellow;

Timer_Process:
  process
  begin
    Timed_Out_Long <= transport '0', '1' after Long_Time;
    Timed_Out_Short <= transport '0', '1' after Short_Time;
    wait on Start_Timer;
  end proccss;

end revised;
```

Notice in this version of the specification body all the timer elements have been changed to the type BIT.

The First Partition

Given the new definition of the body of the specification, it is possible to partition out one of the components. The first component partitioned out is the timer section of the device. The timer section relies on three pieces of information: the values of the Long_Time and Short_Time generics of the device and the Start_Timer signal. This implies that the timer component must have two generics (to pass in the values for Long_Time and Short_Time) and one input signal to start the timer. The timer section produces two results, the Timed_Out_Long signal pulse and the Timed_Out_Short signal pulse. As in the specification, generics will be used to pass in the values for the long and short delays, so we will need generics similar to those used in the interface. The Timer_Process process in the specification is sensitive to the signal Start_Timer and assigns to the signals Timed_Out_Long and Timed_Out_Short. Based on this information, the following entity is created.

```
use work.Design_Package.all;
entity Timer is
  generic (
    Long_Time : Time;
    Short_Time : Time);
  port (
    Start : in Data_Type;
    TL, TS: out Data_Type);
end Timer;
```

To create the architectural body for the timer circuit, the process is taken from the specification and the signal names are replaced with the corresponding port names in the timer interface. The following body is used for the timer circuit:

```
architecture Behavior of Timer is
begin

Timer_Process:
process(Start)
begin
  TL <= transport '0', '1' after Long_Time;
  TS <= transport '0', '1' after Short_Time;
end process;

end Behavior;
```

The new design entity named Timer can now be used in the main circuit to replace the process statement in the specification. The main circuit architectural body is renamed, so that the original body can be retained in the library. Notice that the interface to the main circuit does not need to be changed and, therefore, does not need to be reanalyzed.

```
architecture structure_1 of traffic_light_controller is
  signal Present_State : State := Highway_Light_Green;
  signal Timed_Out_Long, Timed_Out_Short : Data_Type := '0';
  signal Start_Timer : Data_Type := '0';

  component Timer_Section
    generic (Long_Time, Short_Time : Time);
    port (Start : in Data_Type; TL, TS: out Data_Type);
  end component;

begin

  Controller_Process:
  process
  begin
    case Present_State is
      when Highway_Light_Green =>
        if Car_On_Farmroad and Timed_Out_Long = '1' then
          Start_Timer <= transport not Start_Timer;
          Present_State <= transport Highway_Light_Yellow;
        end if;
      when Highway_Light_Yellow =>
        if Timed_Out_Short = '1' then
          Start_Timer <= transport not Start_Timer;
          Present_State <= transport Farmroad_Light_Green;
        end if;
      when Farmroad_Light_Green =>
        if not Car_On_Farmroad or Timed_Out_Long = '1' then
          Start_Timer <= transport not Start_Timer;
          Present_State <= transport Farmroad_Light_Yellow;
        end if;
      when Farmroad_Light_Yellow =>
        if Timed_Out_Short = '1' then
          Start_Timer <= transport not Start_Timer;
          Present_State <= transport Highway_Light_Green;
        end if;
    end case;
    wait on
      Car_On_Farmroad, Timed_Out_Long, Timed_Out_Short;
  end process;
```

```
Highway_Light_Set:
  with Present_State select
    Highway_Light <=
      Green when Highway_Light_Green,
      Yellow when Highway_Light_Yellow,
      Red when
        Farmroad_Light_Green I Farmroad_Light_Yellow;

Farmroad_Light_Set:
  with Present_State select
    Farmroad_Light <=
      Green when Farmroad_Light_Green,
      Yellow when Farmroad_Light_Yellow,
      Red when
        Highway_Light_Green I Highway_Light_Yellow;

Timer_Struct : Timer_Section
  generic map (
    Long_Time,
    Short_Time)
  port map (
    Start_Timer,
    Timed_Out_Long,
    Timed_Out_Short);

end structure_1;
```

Finally, a configuration declaration is written to bind the new timer circuit to the body of the main circuit.

```
use work.all;
configuration struct1 of TLC_TEST is
  for Test
    for Controller : TLC use
      entity work.Traffic_Light_Controller(Structure_1);
      for Structure_1
        for Timer_Struct : Timer_Section use
          entity work.Timer(Behavior);
        end for;
      end for;
    end for;
  end for;
end struct1;
```

The reconfigured model can now be resimulated to check that the results match those which were generated with the specification.

The Second Partition

Now the circuit can be further partitioned by extracting the controller element itself into a separate component. The first step is to isolate the controller behavior into a separate design unit which is then instantiated from the main circuit.

The controller section makes use of a number of signals. First, since it is the main functional body of the entire circuit, it relies on the interface to the circuit. The controller section must respond to the detection of a car on the farm road. Therefore, an input is needed to pass in the signal Car_On_Farmroad. The controller section will determine the signals for both traffic lights so it will need two outputs for each of the lights. The controller section also initiates the timer with the Start_Timer signal and responds to the two timed outputs of the timer section. Putting this all together, there are three inputs and three outputs. The names used for these inputs and outputs will be less mnemonic because the application of the circuit no longer affects the implementation. The car detection signal will be called C, the highway lights will be called HL and FL (for highway light and farm road light, respectively), the signal to start the timer will be called ST, and the two timeout signals will be passed in through the ports TL and TS (for timed out long and timed out short).

```
use work.traffic_package.all;
use work.design_package.all;
entity TL_Controller is
  port (
    C  : in Data_Type;    -- Car Detected on Farm road
    TL : in Data_Type;    -- Green Light Timed Out
    TS : in Data_Type;    -- Yellow Light Timed Out
    HL : out Color_Bits; -- Highway Light
    FL : out Color_Bits; -- Farm road Light
    ST : out Data_Type); -- Start Timer
  end TL_Controller;
```

The HL and FL ports are declared to be of type Color_Bits because from this point on only bit representations will be used.

The behavior of the controller circuit can now be extracted from the main circuit and placed in a behavioral architectural body of the new entity.

```
architecture Behavior of TL_Controller is
  signal Controller_State : State_Bits := Highway_Light_Green;
begin
    Controller_Process:
    process
      variable Last_ST : Data_Type := '0';
    begin
      case Controller_State is
        when HG_State =>
          if C = '1' and TL = '1' then
            Last_St := not Last_St;
            ST <= transport Last_St;
            Controller_State <= transport HY_State;
          end if;
        when HY_State =>
          if TS = '1' then
            Last_St := not Last_St;
            ST <= transport Last_St;
            Controller_State <= transport FG_State;
          end if;
        when FG_State =>
          if C = '0' or TL = '1' then
            Last_St := not Last_St;
            ST <= transport Last_St;
            Controller_State <= transport FY_State;
          end if;
        when FY_State =>
          if TS = '1' then
            Last_St := not Last_St;
            ST <= transport Last_St;
            Controller_State <= transport HG_State;
          end if;
      end case;
      wait on C, TL, TS;
    end process;

  Highway_Light:
    with Controller_State select
      HL <=
        Green_Light when HG_State,
        Yellow_Light when HY_State,
        Red_Light when FG_State | FY_State;

  Farmroad_Light:
    with Controller_State select
      FL <=
        Green_Light when FG_State,
```

```
        Yellow_Light when FY_State,
        Red_Light when HG_State I HY_State;

  end Behavior;
```

The main design unit is then replaced with a new architectural body which instantiates the new controller circuit as well as the timer circuit developed above. The new architectural body is given below.

```
    use work.design_package.all;
    architecture structure_2 of traffic_light_controller is

      signal Timed_Out_Long, Timed_Out_Short : Data_Type := '0';
      signal Start_Timer : Data_Type := '0';

      component Timer_Section
        generic (Long_Time, Short_Time : Time);
        port (Start : in Data_Type; TL, TS: out Data_Type);
      end component;

      component Controller_Section
        port (C, TL, TS : in Data_Type;
              HL, FL : out Color_Bits;
              ST: out Data_Type);
      end component;

    begin

      Traffic_Light: Controller_Section
        port map (
          Bit_To_Boolean(Car_On_Farmroad),
          Timed_Out_Long,
          Timed_Out_Short,
          Bits_To_Color(HL) => Highway_Light,
          Bits_To_Color(FL) => Farmroad_Light,
          ST => Start_Timer);

      Timer_Struct : Timer_Section
        generic map (
          Long_Time,
          Short_Time)
        port map (
          Start_Timer,
          Timed_Out_Long,
          Timed_Out_Short);
```

end structure_2;

The following configuration is used to configure the circuit with the new component.

```
use work.all;
configuration struct2 of TLC_TEST is
  for Test
    for Controller : TLC use
      entity Traffic_Light_Controller(Structure_2);
      for Structure_2
        use work.all;
        for Traffic_Light : Controller_Section use
          entity work.tl_controller(behavior);
        end for;
        for Timer_Struct : Timer_Section use
          entity work.timer(behavior);
        end for;
      end for;
    end for;
  end for;
end struct2;
```

The circuit can now be resimulated to check that the main circuit still exhibits the same behavior.

Starting the Implementation

The traffic light controller circuit has now been partitioned into two components. The VHDL description can now be given to each team to refine into an implementation. Each team will be able to test (simulate) their design at each point of the development, regardless of the progress of the other team. This section will follow the design of the team working on the controller section.

Now that the controller section has been partitioned, some additional design decisions will be made. The implementation considered here is a PLA circuit. This implies that the functionality of the circuit should be broken down into its sum of products terms. One way to do this is to first convert the transition table for the circuit into a bit level representation. The extended table is given in Figure 8.5.

Based on this state table we can then develop a different architectural body for tl_controller which describes the behavior of the

Inputs			Present State		Next State			Outputs			
C	TL	TS	Y(0)	Y(1)	Y(0)	Y(1)	ST	HL(0)	HL(1)	FL(0)	FL(1)
0	X	X	0	0	0	0	0	0	0	1	0
X	0	X	0	0	0	0	0	0	0	1	0
1	1	X	0	0	0	1	1	0	0	1	0
X	X	0	0	1	0	1	0	0	1	1	0
X	X	1	0	1	1	1	1	0	1	1	0
1	0	X	1	1	1	1	0	1	0	0	0
0	X	X	1	1	1	0	1	1	0	0	0
X	1	X	1	1	1	0	1	1	0	0	0
X	X	0	1	0	1	0	0	1	0	0	1
X	X	1	1	0	0	0	1	1	0	0	1

Figure 8.5. Extended Transition Table for Traffic Light Controller

circuit in the terms of the state table. The state table representation is more obscure than the general behavioral description, so it is useful to keep the description of the general behavior to document the intentions of the state table.

```
architecture Behavior_2 of TL_Controller is
  signal Controller_State : State := Highway_Light_Green;
  signal Y : State_Bits := B"00";
begin
  Y <= State_To_Bits(Controller_State);

process
  variable SumOfProducts : Bit_Vector(0 to 6);
begin
  Controller_State <= Bits_To_State(Y);
  if   C = '0' and Y(0) = '0' and Y(1) = '0' then
    SumOfProducts := B"0000010";
  elsif TL = '0' and Y(0) = '0' and Y(1) = '0' then
    SumOfProducts := B"0000010";
  elsif C= '1' and TL= '1' and Y(0)= '0' and Y(1)= '0' then
    SumOfProducts := B"0110010";
  elsif TS = '0' and Y(0) = '0' and Y(1) = '1' then
    SumOfProducts := B"0100110";
  elsif TS = '1' and Y(0) = '0' and Y(1) = '1' then
    SumOfProducts := B"1110110";
  elsif C= '1' and TL= '0' and Y(0)= '1' and Y(1)= '1' then
    SumOfProducts := B"1101000";
  elsif C = '0' and Y(0) = '1' and Y(1) = '1' then
```

```
              SumOfProducts := B"1011000";
            elsif TL = '1' and Y(0) = '1' and Y(1) = '1' then
              SumOfProducts := B"1011000";
            elsif TS = '0' and Y(0) = '1' and Y(1) = '0' then
              SumOfProducts := B"1001001";
            elsif TS = '1' and Y(0) = '1' and Y(1) = '0' then
              SumOfProducts := B"0011001";
            end if;
            Controller_State <= transport
              Bits_to_State(State_Bits(SumOfProducts(0 to 1)));
            ST <= transport SumOfProducts(2);
            HL <= transport Color_Bits(SumOfProducts(3 to 4));
            FL <= transport Color_Bits(SumOfProducts(5 to 6));
            wait on C, TL, TS, Y(0), Y(1);
          end process;
        end Behavior_2;
```

Again, a new configuration declaration is written to bind the new component into the circuit.

```
        use work.all;
        configuration struct3 of TLC_TEST is
          for Test
            for Controller : TLC use
              entity Traffic_Light_Controller(Structure_2);
              for Structure_2
                use work.all;
                for Traffic_Light : Controller_Section use
                  entity work.tl_controller(behavior_2);
                end for;
                for Timer_Struct : Timer_Section use
                  entity work.timer(behavior);
                end for;
              end for;
            end for;
          end for;
        end struct3;
```

Again, the model can be simulated to check that the results match those generated in previous simulations.

Setting Up the PLA

The PLA section of the circuit can now be developed. Therefore, the functionality of the circuit should be described in sum of products terms. These terms can be generated directly from the transition table. The first step is to develop a behavioral model of the PLA section.

From the state table, it can be seen that the PLA circuit will have 5 inputs and 7 outputs. The following entity declaration will be used for the PLA circuit.

```
entity PLA is
  port (In0, In1, In2, In3, In4: in Bit;
        Out0, Out1, Out2, Out3, Out4, Out5, Out6: out Bit);
end PLA;
```

The transition table is then used to directly generate the sum of products terms for the circuit.

```
architecture Behavior1 of PLA is
  type PLA_Matrix is array
    (Integer range 0 to 9, Integer range 0 to 6) of Bit;
  constant PLA_Outputs : PLA_Matrix := (
    ('0', '0', '0', '0', '0', '1', '0'),
    ('0', '0', '0', '0', '0', '1', '0'),
    ('0', '1', '1', '0', '0', '1', '0'),
    ('0', '1', '0', '0', '1', '1', '0'),
    ('1', '1', '1', '0', '1', '1', '0'),
    ('1', '1', '0', '1', '0', '0', '0'),
    ('1', '0', '1', '1', '0', '0', '0'),
    ('1', '0', '1', '1', '0', '0', '0'),
    ('1', '0', '0', '1', '0', '0', '1'),
    ('0', '0', '1', '1', '0', '0', '1')));

begin

process
  variable New_State : Integer;
begin
  if In0= '0' and In3= '0' and In4= '0' then
    New_State := 0;
  elsif In1= '0' and In3= '0' and In4= '0' then
    New_State := 1;
  elsif In0= '1' and In1= '1' and In3= '0' and In4= '0' then
    New_State := 2;
```

```
    elsif In2= '0' and In3= '0' and In4= '1' then
      New_State := 3;
    elsif In2= '1' and In3= '0' and In4= '1' then
      New_State := 4;
    elsif In0= '1' and In1= '0' and In3= '1' and In4= '1' then
      New_State := 5;
    elsif In0= '0' and In1= '0' and In3= '1' and In4= '1' then
      New_State := 6;
    elsif In1= '1' and In3= '1' and In4= '1' then
      New_State := 7;
    elsif In2= '0' and In3= '1' and In4= '0' then
      New_State := 8;
    elsif In2= '1' and In3= '1' and In4= '1' then
      New_State := 9;
    else
      assert (FALSE) report "Error in PLA" severity Error;
    end if;

    Out0 <= PLA_Outputs(New_State, 0);
    Out1 <= PLA_Outputs(New_State, 1);
    Out2 <= PLA_Outputs(New_State, 2);
    Out3 <= PLA_Outputs(New_State, 3);
    Out4 <= PLA_Outputs(New_State, 4);
    Out5 <= PLA_Outputs(New_State, 5);
    Out6 <= PLA_Outputs(New_State, 6);

    wait on In0, In1, In2, In3, In4;
  end process;

end Behavior1;
```

A new architectural body is created for the controller circuit to replace the behavioral code with an instantiation of the PLA behavior.

```
architecture Structure of TL_Controller is
    signal Controller_State : State := Highway_Light_Green;
    signal YIn : State_Bits := B"00";
    signal YOut : State_Bits := B"00";
    component PLA_Section
      port (In0, In1, In2, In3, In4 : in Bit;
            Out0, Out1, Out2, Out3, Out4, Out5, Out6 : out Bit);
    end component;
begin
    YIn <= State_To_Bits(Controller_State);
    Controller_State <= Bits_To_State(YOut);
```

```
      Impl : PLA_Section port map
      (C, TL, TS, YIn(0), YIn(1),
        YOut(0), Yout(1), ST, HL(0), HL(1), FL(0), FL(1));
  end Structure;
```

The following configuration is used to configure the new circuit.

```
  use work.all;
  configuration struct4 of TLC_TEST is
    for Test
      for Controller : TLC use
        entity Traffic_Light_Controller(Structure_2);
        for Structure_2
          use work.all;
          for Traffic_Light : Controller_Section use
            entity work.tl_controller(structure);
            for Structure
              for Impl: PLA_Section use
                entity work.pla(behavior1);
              end for;
            end for;
          end for;
          for Timer_Struct : Timer_Section use
            entity work.timer(behavior);
          end for;
        end for;
      end for;
    end for;
  end struct4;
```

The model can now be simulated again.

The reader is left to further develop the component sections of the PLA circuit and the circuit for the timer model.

This comprehensive example is included here to illustrate the use of VHDL throughout the design process. A high-level specification of a desired circuit was developed and test input was written to stimulate the circuit. The circuit was then partitioned into two sections, a timer section and a controller section. The controller section was successively refined into a PLA representation. The important point to note here is that at each level of refinement the circuit could be simulated and checked. This method of working allows a designer to catch design errors or inconsistencies early in the design process.

This chapter has brought together many of the elements of VHDL described in the previous chapters. The main elements of VHDL have been presented and some indications of their use have been examined. The next chapter will look at some of the more advanced features of VHDL. This will be followed by some smaller examples of using VHDL in other models, particularly in modeling systems.

Chapter 9
Advanced Features

This chapter presents several advanced features provided by VHDL. Discussed in the following sections are:

Overloading
Access Types
File Types and I/O
User-Defined Attributes
Signal-Related Attributes
Aliases
Association by Subelement
Guarded Signal Assignment Statements
Disconnection Specifications
Null Transactions

Overloading

An item is overloaded if it has been defined to have more than one meaning and if the particular meaning of a use of the item is determined by type information in the local context of use. In VHDL, enumeration literals (character literals or identifiers), the names of procedures and functions, and operator symbols (like "+", "**not**", and "<") may be overloaded. Overloading is a useful feature for a hardware description language since it allows a designer to capture the fact that an operation defined for objects of one type is at some level the same as an operation defined for objects of a second type.

The simplest kind of overloading is defining a character literal or identifier to be an enumeration literal of two different enumeration types. The predefined VHDL package Standard contains an example of overloading: the character literals '0' and '1' are defined to be simultaneously values of type Character and values of type Bit. Literals like '0' and '1' and identifiers like High and Low are conventionally used to represent a number of different data types; overloading allows the designer to use the conventional designations for values while providing a mechanism for disambiguating each use.

```
type Four_level is (Rising, High, Falling, Low) ;
type Two_level is (High, Low) ;
   -- High and Low are overloaded

signal S1 : Four_level := Low ;
signal S2 : Two_level := Low ;
   -- Different meanings of Low, disambiguated by type
   -- information in the context of a signal declaration

type Three_value_logic is ('0', '1', 'X') ;
type Four_value_logic is ('0', '1', 'Z', 'X') ;
   -- '0', '1', and 'X' are overloaded

signal S3 : Three_value_logic := 'X' ;
signal S4 : Four_value_logic := 'X' ;
   -- Different meanings of 'X', disambiguated by type
   -- information in the context of a signal declaration
```

Names of procedures and functions may be overloaded provided the profiles of their parameter types (and return type in the case of functions) are different. In the following example, the names Invert and Transform are overloaded. Because the profiles of their parameter types are different, uses of the names can be disambiguated by type information in the context of use.

```
function Invert (X : Three_level_logic)
   return Three_level_logic ;
function Invert (X : Matrix) return Matrix ;
procedure Invert (X : inout Three_level_logic) ;
procedure Invert (X : inout Matrix) ;
function Transform (C : Cartesian) return Polar ;
function Transform (P : Polar) return Cartesian ;
```

Note that it is the types of the parameters, not their names, that must be different. The following VHDL is incorrect since it attempts to overload a name by defining two procedures having different parameter names but having the same parameter type profiles.

> **procedure** Transpose (
>   Cell_1 : **inout** Integer ;
>   Cell_2 : **inout** Integer) ;
>
> **procedure** Transpose (
>   X_coord : **inout** Integer ;
>   Y_coord : **inout** Integer) ;

Operator symbols like "+", "**not**" and "<" may be overloaded as functions but not as procedures. If the operator is a binary operator, then the overloaded function must have two parameters; if the operator is a unary operator, then the overloaded function must have one parameter. In the function specification, the operator symbol must be given in quotation marks. Two styles of function calls are possible: normal function notation (but with the name of the operator in quotation marks) or infix notation.

> **type** Twos_comp **is array** (Natural **range** <>) **of** Bit ;
> **function** "<" (L, R: Twos_comp) **return** Boolean ;
>
> Z <= "<" (RegA, RegB) ;
> Z <= RegA < RegB ;

The statement that overloading can be disambiguated by type information in the context of use is actually a bit oversimplified. In some contexts, an expression involving an overloaded item (typically an enumeration literal) may need to be *qualified* so that its meaning may be unambiguous. An expression is qualified by enclosing the expression in parentheses and prefixing the parenthesized expression with the name of a type and a tic ('). Consider the following the following procedures declarations:

> **procedure** To_natural (X : Character ; Y : **inout** Natural) ;
> **procedure** To_natural (X : Bit ; Y : **inout** Natural) ;

Since the character literal '1' is overloaded, being a value of the type Character and of the type Bit, the procedure call

To_natural ('1', N) ;

would be ambiguous. Qualifying the first actual with the name of the type clears the ambiguity.

To_natural (Character'('1'), N) ;
To_natural (Bit'('1'), N) ;

Access Types

Dynamic memory allocation is a useful capability when doing high-level behavioral modeling and when modeling hybrid hardware/software systems. Dynamic memory allocation requires access (or pointer) types.

An access type is a type whose values are pointers to (or links to, or addresses of) dynamically-allocated, and therefore unnamed, objects of some other type. A value of an access type results from evaluation of a special kind of expression called an allocator. Evaluation of an allocator creates an object which has no simple name and returns an access value which designates this object. The access value can be stored in an object of the appropriate access type. The reserved word **null** is used to represent the null access value, that is, an access value that designates no object. The default initial value of any access object is **null**.

There are two types of allocators. The first kind of allocator consists of the reserved word **new** followed by the name of a type or subtype and possibly a constraint on the type or subtype. This kind of allocator creates an object of the designated type and initializes the object to the default initial value for its type (T'LEFT for a scalar type T or for each scalar element of a composite type). The second kind of allocator consists of the reserved word **new** followed by the name of a type or subtype, the tick (') token, and a parenthesized value of the named type or subtype. This kind of allocator creates an object with an explicit initial value.

```
type Int is range 0 to 500 ;

type Int_ptr is access Int ;
  -- Int_ptr is an access type

variable V_int_ptr_1 : Int_ptr := new Int ' (362) ;
  -- V_int_ptr_1 is a variable of an access type.
```

-- The value of variable V_int_ptr_1 is an access value
--   returned by the allocator.
-- The allocator creates an object whose initial value is 362

variable V_int_ptr_2 : Int_ptr := **new** Int ;
-- The allocator creates an object with
--   the initial value 0 (Int'LEFT)

variable V_int_ptr_3 : Int_ptr ;
-- The initial value of V_int_ptr_3 is **null**.

Appending .**all** to the name of an object of an access type results in a reference to the value of the object designated by the value of the access object. Thus in the above example, the value of the expression V_int_ptr_1.**all** would be 362. If the access object is an array type, then an indexed name consisting of the name of the access object followed by a parenthesized list of index values will reference an element of the object designated by the value of the access object. Likewise, if the access object is a record type, then a selected name consisting of the name of the access object followed by a dot ('.') and a record field designator will reference the value of a field of the object designated by the value of the access object.

type Matrix **is array** (1 **to** 10, 1 **to** 10) **of** Integer ;
type Matrix_ptr **is access** Matrix ;
variable M : Matrix_ptr :=
    new Matrix ' (**others** => (**others** => 0)) ;
variable Left_corner_cell : Integer := M (1, 1) ;
-- The expression M (1, 1) references an element of the
-- two-dimensional array object designated by the value of
-- the access object M.

type Rational **is record**
  Numerator : Integer ;
  Denominator : Integer ;
end record ;
type Rational_ptr **is access** Rational ;
variable R : Rational_ptr := **new** Rational ' (5, 7) ;
variable Num : Integer := R.Numerator ;
-- The expression R.Numerator references the Numerator field
-- of the record object designated by the value of the
-- access object R.

If the value of an access object A is **null**, then any attempt to reference the object designated by the value of A is an error. Thus an expression like A.**all**, A(1) or A.F would be in error if the value of A were **null**.

For each access type A, a deallocation procedure is automatically defined.

procedure Deallocate (P : **inout** A) ;

Deallocate is invoked to indicate that the object designated by the value of the access object P will not be referenced in the future. The value of P is set to **null**. Note that it is the responsibility of the designer to insure that a description does not attempt to reference a deallocated object via "dangling references".

```
type Int is range 0 to 500 ;
type Int_ptr is access Int ;
variable V1 : Int_ptr := new Int ' (100) ;
variable V2 : Int_ptr := V1 ;
   -- The values of V1 and V2 designate the same object
variable V  : Int ;
...
Deallocate (V1) ;
...
V := V2.all ;
   -- V2 is a dangling reference, and
   -- there is no guarantee that V2 still points to the same object
```

VHDL provides an incomplete type declaration that makes it possible to declare the interdependent pointer-and-cell types typically used to describe linked data structures. An incomplete type declaration simply names a type without giving any details; this makes the type name available for use in other type definitions. The full type declaration is given later. The following example illustrates a package declaring a stack data type and operations on this type. It illustrates the use of the incomplete type declaration and most of the other features discussed in this section.

```
package Stack_package is

type Element ;
type ElementPtr is access Element ;
type Element is record
  Value : Integer ;
```

```
      NextElement : ElementPtr ;
   end record ;

   type Stack ;
   type StackPtr is access Stack ;
   type Stack is record
     StackElement : ElementPtr ;
     NumberOfElements : Integer ;
   end record ;

   function CreateStack return StackPtr ;
   procedure Push (
      TheStack : inout StackPtr ;
      Value : in Integer) ;
   procedure Pop (
      TheStack : inout StackPtr ;
      Value : out Integer) ;
   procedure DeleteStack (
      TheStack : inout StackPtr) ;

end Stack_package ;

package body Stack_package is

   function CreateStack return StackPtr is
   begin
     return new Stack ' (null, 0) ;
   end CreateStack ;

   procedure Push (
      TheStack : inout StackPtr ;
      Value : in Integer)
   is
   begin
     TheStack.StackElement := new Element '
     (Value, TheStack.StackElement) ;
     TheStack.NumberOfElements :=
         TheStack.NumberOfElements + 1 ;
   end Push ;

   procedure Pop (
      TheStack : inout StackPtr ;
      Value : out Integer)
   is
    variable OldElement : ElementPtr := TheStack.StackElement;
   begin
    Value := OldElement.Value;
```

```
         TheStack.StackElement := OldElement.NextElement;
         TheStack.NumberOfElements := TheStack.NumberOfElements -1;
         Deallocate (OldElement) ;
      end Pop ;

      procedure DeleteStack (
         TheStack : inout StackPtr)
      is
         variable AValue : Integer ;
      begin
         while TheStack.StackElement /= null loop
            Pop (TheStack, AValue) ;
         end loop ;
      end DeleteStack ;

   end Stack_package ;
```

File Types and I/O

File types and file objects provide a way for a VHDL design to communicate with an external design environment. A VHDL design can write values to any number of files and read values from any number of files. A file is always a file of a particular file type which can only hold values of a particular type. File types are declared in file type declarations.

```
      type Integer_file is file of Integer ;
      type Signal_stream is file of Bit_vector(7 downto 0) ;
      type Bit_history is file of Bit ;
```

It is not possible to declare file types that hold multi-dimensional array types, access types, or file types.

For each file type declaration, VHDL provides three implicitly declared subprogram declarations. Given the following file type declaration,

```
      type D is file of T ;
```

the following subprograms are implicitly declared (that is, they are available for use without the designer having to declare them explicitly):

```
procedure Read (F : in D ; Value : out T) ;
  -- Reads the next value from file F
procedure Write (F : out D ; Value : in T) ;
  -- Appends a value to file D
function Endfile (F : in D) return Boolean ;
  -- Returns true if there are no more values to be
  --   read in file D
```

In case the file is a file of an unconstrained array type, the implicitly declared Read program is of the form:

```
procedure Read (
  F : in D ;
  Value : out T ;
  Length : out Natural) ;

    -- Places the number of the array elements read in
    -- parameter Length; if parameter Value is not sufficiently
    -- long to hold all the elements, data will be lost
```

A file object declaration has a type, a mode (**in** or **out**; files of mode **inout** are not supported), and a string expression naming the file. The contents of the string expression are interpreted by the external environment.

The following example shows how to record a signal history in a signal trace file and how to play this signal history back into a design entity.

```
package Signal_trace_file is
  type Waveform_element is record
    Value : Bit ;
    At    : Time ;
  end record ;
  type Signal_history is file of Waveform_element ;
end Signal_trace_file ;

use Work.Signal_trace_file.all ;
entity Signal_recorder is
end Signal_recorder ;

architecture Signal_recorder of Signal_recorder is
  file STF : Signal_history is out "Stream_1" ;
  signal S : Bit := '0' ;
begin
  S <=
```

```
             '1' after 10 ns ,
             '0' after 20 ns ,
             '1' after 30 ns ,
             '0' after 40 ns ,
             '1' after 50 ns ,
             '0' after 60 ns ,
             '1' after 70 ns ,
             '0' after 80 ns ,
             '1' after 90 ns ,
             '0' after 100 ns ;
           process
             variable V : Waveform_element ;
           begin
             V.Value := S ;
             V.At := Now ;
             Write (STF, V) ;
             wait on S ;
           end process ;
         end Signal_recorder ;

         use Work.Signal_trace_file.all ;
         entity Signal_playback is
         end Signal_playback ;

         architecture Signal_playback of Signal_playback is
           file STF : Signal_history is in "Stream_1" ;
           signal S : Bit := '0' ;
         begin
           process
             variable V : Waveform_element := ('0', Now) ;
           begin
             S <= V.Value ;
             if Endfile (STF) then
               wait ;
             else
               Read (STF, V) ;
               wait for V.At - Now ;
             end if ;
           end process ;
         end Signal_playback ;
```

User-Defined Attributes

While VHDL is a rich language, there will always be some information about a particular design which cannot be directly described using the language constructs available. For instance, VHDL does not provide direct mechanisms for describing the physical characteristics of a design such as placement for layout or wire widths. VHDL does provide the means, however, for a designer or design team to incorporate additional information within a VHDL description through the use of user-defined attributes.

A user-defined attribute can be any type of information which needs to be associated with an element of a VHDL description. An attribute can be associated with any of the following: entities, architectures, configurations, subprograms, packages, types and subtypes, constants, signals, variables, components, or labels. User-defined attributes must be constants. Attributes can be of any type except an access or file type. This flexibility allows a designer plenty of freedom for annotating a VHDL description with all types of information which do not fall naturally into the language itself.

In order to use an attribute it is necessary to define it using an attribute declaration. the attribute declaration has the following form:

> **attribute** *identifier* : *type-mark*;

The following package illustrates the use of attribute declarations. The package includes some types used for physical layout and defines attributes which will be applied to the elements which affect the layout of a design.

```
package Physical_Attributes is

   type Physical_Size is record
      Width, Height : Integer;
   end record;

   type Location is record
      X, Y : Integer;
   end record;

   attribute LayoutSize : Physical_Size;
   attribute Placement : Location;
   attribute Pin_Number : Integer;
   attribute Width : Integer;
```

end Physical_Attributes;

To associate an attribute with an object or class of objects an attribute specification is used. The attribute specification has the form:

attribute *attribute-designator*
of *entity-specification*
is *expression*;

The following architecture associates physical attributes with elements of the description.

architecture NAND_Impl **of** SR_Flip_Flop **is**
component NAND_Gate
port (A, B : **in** Bit; Y : **out** Bit);
end component;
attribute Placement **of** NAND1 : **label is** (200, 100);
attribute Placement **of** NAND2 : **label is** (200, 90);
begin
NAND1 : NAND_Gate **port map** (S, QBar, Q);
NAND2 : NAND_Gate **port map** (S, Q, QBar);
end NAND_Impl;

Once an attribute has been defined and associated with an object, an attribute name is used to reference the value of the attribute for that object.

prefix' attribute-designator

For example, to read the Placement attribute of the component label NAND1 above, the following expression is used:

NAND1'Placement

User-defined attributes provide a powerful means for extending VHDL to meet the needs of a specific project or application. They can be used to annotate the design with additional information and provide a means of giving information to other CAE tools.

Signal-Related Attributes

VHDL contains a number of predefined attributes which are related to signals. These can be divided into two classes: those attribute which define signals themselves ('Delayed, 'Stable, 'Quiet, and 'Transaction) and those attributes which are functions that provide information about signals ('Active, 'Event, 'Last_Value, 'Last_Event, and 'Last_Active). This section examines each of these attributes and shows examples of where they might be used.

The 'Delayed attribute echoes the value of the prefix signal, delayed by the specified time factor. For instance, given the following signal declaration and assignment

> **signal** S : Integer;
> S <= 1 **after** 1 ns, 2 **after** 2 ns;

the value of S'Delayed(10 ns) will be 1 after 11 ns and 2 after 12 ns. The 'Delayed attribute might be used in a phase-shifted clock as in the following design unit:

> **entity** Clock **is**
>   **generic** (CycleTime : Time := 25 ns);
>   **port** (Phase0, Phase1: **out** Bit);
> **end** Clock;
>
> **architecture** UsingAttributes **of** Clock **is**
>
>   **signal** ControlSignal : Bit := '0';
>
> **begin**
>
>   ControlSignal <= **not** ControlSignal **after** CycleTime;
>
>   Phase0 <= ControlSignal;
>   Phase1 <= ControlSignal'Delayed(CycleTime/2);
>
>   **end** UsingAttributes;

Phase1 of the clock is set by delaying the value of the control signal by half the cycle time.

The 'Quiet attribute is a boolean signal whose value is TRUE when a transaction has not occurred on the signal for the given time; otherwise, it is FALSE. A transaction is any assignment to the signal, including an assignment of the same value as the previous value of the signal. The 'Stable attribute is a boolean signal whose value is TRUE when an event has not occurred on the signal for the given time; otherwise, it is FALSE. These signal-valued attributes are useful for many situations, including the detection of spikes on inputs lines as in the following design example:

```
entity DFF is
  port (
    D : in Bit;
    Q : out Bit);
end DFF;

architecture Spike_Protected of DFF is
begin
  process
  begin
    wait until D'Stable(5 ns);
    Q <= D after 7 ns;
  end process;
end Spike_Protected;
```

The signal D'Stable(5 ns) goes to FALSE every time D changes and returns to TRUE when D has not changed for 5 ns. Therefore, the flip flop will not activate until the input D has remained stable for 5 ns.

The 'Transaction attribute is a bit signal which toggles every time the signal becomes active.

For instance, suppose a process is set up to count the number of outputs from a component. A VHDL process could be set up which tracks the transactions on the output of the component and increments a counter.

```
entity NandComponent is
  port (I0, I1 : in Bit; Output : out Bit);
end NandComponent;

architecture Accounting of NandComponent is

  component Nand_Gate
    port (A, B : in Bit; Y : out Bit);
  end component;

  signal Result : Bit;
```

```
begin

   L1 : Nand_Gate
     port map (In0, In1, Result);
   Output <= Result;

   process
     variable Count : Integer := 0;
   begin
     wait on Result'Transaction;
     Count := Count + 1;
   end process;

end Accounting;
```

In this example, the two input ports of the design entity are passed into the Nand_Gate component. A Result signal is used as the output of the component so that the architecture can read the result before passing it through the output port. The process statement is sensitive to Result'Transaction since the process is counting activations of the Nand_Gate, not changes on the output of the Nand_Gate.

The attributes which are functions returning information about signals are 'Active, 'Event, 'Last_Value, 'Last_Active, and 'Last_Event. They are used when information about the status of a signal is needed but the process does not need to be sensitive to that information directly.

The 'Active attribute is TRUE when a transaction has occurred on the signal during the current simulation cycle. The 'Event attribute is TRUE when an event has occurred on the signal during the current simulation cycle. For example, the following process is sensitive to two input signals but its behavior depends on whether or not each of the inputs has changed.

```
entity Buffer is
  port (
    In1, In2 : Bit;
    Output : out Bit);
end Buffer;

architecture Non_Concurrent of Buffer is
begin
  process
  begin
    wait on In1, In2;
    assert In1'Active xor In2'Active
      report "Concurrent Input to Buffer - Flowthrough Blocked"
```

```
        severity Warning;
     if In1'Active then
       Output <= In1;
     else
       Output <= In2;
     end if;
   end process;
 end Non_Concurrent;
```

The attribute 'Last_Active returns the amount of time that has elapsed since there was a transaction on the signal and 'Last_Event return the amount of time that has elapsed since there was an event on the signal. The 'Last_Value attribute returns the value of the signal before the last event. This last attribute is used in determining the previous state of a signal, as in the following example:

```
entity Shift_Register is
  port (
    InBit, Clock : in Bit;
    OutBit : out Bit);
end Shift_Register;

architecture Falling_Edge of Shift_Register is
begin
  process
    variable Register : Bit_Vector(7 downto 0) := "00000000";
  begin
    if Clock'Last_Value = '1' and Clock = '0' then
      OutBit <= Register(0);
      for I in 0 to 6 loop
        Register(I) := Register(I+1);
      end loop;
      Register(7) := InBit;
    end if;
    wait on Clock;
  end process;
end Falling_Edge;
```

Aliases

An alias is an alternate designation for an object. The general form of the alias declaration is

> **alias** *identifier* : *subtype_indication* **is** *name*

A *subtype_indication* is the name of a type (or subtype) with an optional range constraint or index constraint. The fact that a constraint may appear on the alias means that the subtype of the alias may be different from the subtype of the aliased name; however, their types must be the same. The type of the alias may not be a multi-dimensional array. Aliasing is especially useful for giving a simple name to a subelement (or collection of subelements) of a composite object.

> **signal** Instruction : Bit_vector (15 **downto** 0) ;
> **alias** Opcode : Bit_vector (3 **downto** 0) **is**
>   Instruction (15 **downto** 12) ;
> **alias** Operand_1 : Bit_vector (5 **downto** 0) **is**
>   Instruction (11 **downto** 6) ;
> **alias** Operand_2 : Bit_vector (5 **downto** 0) **is**
>   Instruction (5 **downto** 0) ;
> **alias** Indirect_bit : Bit **is** Opcode (12) ;
>   -- An alias of an alias
>
> **variable** Condition_code : Bit_vector (3 **downto** 0) ;
> **alias** Overflow : Bit **is** Condition_code (2) ;

An alias of an object can be updated if and only if it is an alternate name for an object which can be updated. Thus an alias of a constant or a port of mode **in** cannot be updated.

Association by Subelement

Subprogram calls, component instantiation statements, and the binding indications of configuration specifications all involve associating actual objects or values with formal objects (subprogram parameters, ports, and generics). In an association, the formal object is normally considered as a whole, even if it is a composite (array or record); that is, an actual is normally associated with the whole formal, not with a part of the formal. However, there are circumstances where it is useful to be able to associate an actual with a subelement of a formal. The following example illustrates such a case. The entity Mux has a port Sel of type Two_bit, which is a descending array of two elements. The architecture Mux_user instantiates Mux and associates, via a binding indication in the configuration specification, the signal Temperature_sensor with one subelement of Sel and the signal Air_sensor with the other subelement of Sel.

```
package Types is
  subtype Two_bit is Bit_vector (1 downto 0) ;
  subtype Four_bit is Bit_vector (3 downto 0) ;
end Types ;

use Work.Types.all ;

entity Mux is
  port (Enable : Bit ;
       Sel : Two_bit ;
       Inputs : Four_bit ;
       Output : out Bit) ;
end Mux ;

entity Mux_user is
end Mux_user ;

use Work.Types.all ;
architecture Mux_user of Mux_user is
  component Mux
    port (Enable : Bit ;
         Sel : Two_bit ;
         Inputs : Four_bit ;
         Output : out Bit) ;
  end component ;
  signal Mux_enable : Bit ;
  signal Temperature_sensor : Bit ;
  signal Air_sensor : Bit ;
  signal Four_bus : Four_bit ;
  signal Mux_out : Bit ;
  for M : Mux use entity Work.Mux ;
begin
  M : Mux port map (Enable => Mux_enable,
                  Sel (1) => Temperature_sensor,
                  Sel (0) => Air_sensor,
                  Inputs => Four_bus,
                  Output =>Mux_out) ;
end Mux_user ;
```

Each subelement of a formal must be associated with exactly one actual. Furthermore, all the subelement associations for a single formal must be contiguous; thus, the following association list would be an error:

M : Mux **port map** (Enable => Mux_enable,
                Sel (1) => Temperature_sensor,
                Inputs => Four_bus,
                Sel (0) => Air_sensor,
                Output =>Mux_out) ;

Guarded Assignment Statements

One of the options available in a concurrent signal assignment is the **guarded** option. This option specifies that the signal assignment statement will only execute when the guard condition of the block statement which contains the assignment is true. Block statements and guard expressions were discussed in the Chapter 7. Recall that the guard condition of a block statement represents a signal with the name GUARD. This signal is used to control the execution of the equivalent process statement for the concurrent signal assignment. In particular, a conditional or selected signal assignment which uses the **guarded** option will result in the following type of process.

```
process
begin
  if GUARD then
    if-or-case-statement
  end if;
  wait on (GUARD + signals-in-transform)
end process;
```

As an example, suppose that an AND gate is dependent on an enable signal, EN, with the value '1'. The following architecture would implement the new behavior.

```
architecture Guarded of AND_Gate is
begin
  block (EN = '1')
  begin
    Y <= guarded
      '1' after Delay when A = '1' and B = '1' else
      '0' after Delay;
  end block;
end Guarded;
```

This architecture is equivalent to the following architecture:

```
architecture GuardedEquivalent of AND_Gate is
begin
  block(EN = '1')
  begin
    process
    begin
      if GUARD then
        if A = '1' and B = '1' then
          Y <= '1' after Delay;
        else
          Y <= '0' after Delay;
        end if;
      end if;
      wait on GUARD, A, B;
    end process;
  end block;
end GuardedEquivalent;
```

Disconnection Specifications

Chapter 5 discussed resolved signals and the effects and requirements of the resolution functions associated with them. Recall that the input to the resolution function is the array of values on all drivers of the signal. In some instances this is more than the designer wants; resolution functions are often used to implement a signal which is driven by one of a possible number of sources. To accommodate this need, VHDL includes two ways of *turning off* particular drivers of a resolved signal. This section discusses one of those; the other is discussed in the next section.

VHDL includes a feature to turn off the driver of a guarded resolved signal called the *disconnection specification*. A guarded signal is one in which the reserved word **register** or **bus** is given as the signal kind. The disconnection specification only applies to drivers of guarded signals. If the guarded signal is of the signal kind **register** then the resolution function is not called when all of the drivers of the signal are disconnected. The disconnection specification is used when the designer wants the driver turned off under conditions defined by the guard expression of the enclosing block. The disconnection specification has the form:

disconnect *signal-list* : *type-name* **after** *time-expression*;

If no disconnection specification is given for a guarded signal, an implicit disconnection after 0 ns is assumed.

The specification applies only to signals declared in the *immediately enclosing* block in which the specification is given. For instance, consider a model which increments an integral signal every time an increment event occurs. If a restart event is detected, the signal is reset to 0 and the count continues. The following package is used to define an event type and a resolution function for an array of integers.

```
package Event_Package is

   type Event is (Increment, Restart);

   type IntArray is array (Integer range <>) of Integer;
   function OnlyOne(Input: IntArray) return Integer;
   subtype ResolvedInt is OnlyOne Integer;

end Event_Package;

package body Event_Package is

   function OnlyOne(Input: IntArray) return Integer is
   begin
     return Input(Input'First);
   end OneOnly;

end Event_Package;
```

Using this package, an architecture is defined which implements the desired behavior of the model.

```
use Work.Event_Package.all;
entity Counter is
   port (InputEvent : Event);
end Counter;

architecture EventDriven of Counter is

   signal Count : ResolvedInt register := 0;

   disconnect Count : ResolvedInt after 0 ns;

begin
```

```
          Counter_Block:
          block (InputEvent = Increment)
          begin
            Count <= guarded Count + 1;
          end block;

          Interrupt_Block:
          block (InputEvent = Restart)
          begin
            Count <= guarded 0;
          end block;

       end EventDriven;
```

In this architecture both blocks are guarded by the value of the input signal InputEvent. When InputEvent is equal to Increment the guard expression on the first block is true and the guard on the second block is false. This causes the first assignment to execute and the disconnection specification for the second driver to go into effect; the net result is an increment of the signal Count. When InputEvent is equal to Restart the first guard expression becomes false and the second guard expression becomes true causing the opposite effect; the first driver's disconnection specification goes into effect and the second assignment is executed (resetting the signal Count to 0).

Null Transactions

The designer may choose to explicitly assign a *null transaction* to a guarded resolved signal in a sequential signal assignment statement. This has the effect that the driver defined by that signal assignment will not be used in the resolution of the signal value; *i.e.* that particular driver is turned off. The waveform element for the null transaction has the form:

       null [after *time-expression*]

For instance, consider the following example which might be taken from the VHDL description of a microprocessor. A package is given to declare the type for the stages of the microprocessor and an unconstrained array type and resolution function are defined for use in declaring resolved signals of this new type.

```
type Stage is (FETCH, DECODE, EXECUTE);
type Stage_Array is array (Integer range <>) of Stage;

function Resolve_Stage (
  Input : Stage_Array)
return Stage;

end Processor_Package;

package body Processor_Package is

function Resolve_Stage (
  Input : Stage_Array)
return Stage
is
begin
  assert (Input'BLength > 1)
    report "Overdriven Bus"
      severity Error;
  return Input(Input'First);
end Resolve_Stage;

end Processor_Package;
```

The following architecture uses the declarations in the above package to define a resolved signal which will hold the current state of the microprocessor.

```
architecture Stages of Processor_State is

subtype Resolved_State is Resolve_Stage Stage;
signal TheStage : Resolved State register;

begin

Cycle_Block:
block(BeginCycle = '1')
begin
  process
  begin
    TheStage <=
      FETCH after 20 ns,
      null after 40 ns,
      null after 60 ns;
  end process;
```

```
process
begin
  TheStage <=
    null after 20 ns,
    DECODE after 40 ns,
    null after 60 ns;
end process;

process
begin
  TheStage <=
    null after 20 ns,
    null after 40 ns,
    EXECUTE after 60 ns;
  end process;
end block;

  end Stages;
```

This architecture first defines a subtype which is used to associate the resolution function with a signal of that subtype. The signal TheStage is then defined to be of that subtype. The architecture contains a block which is guarded by the value of a input port BeginCycle. The block contains three processes which define drivers of the signal TheStage. Each driver is turned off for all but one of the delay time in the waveform. The signal TheStage gets resolved three times in each cycle. The resolution function will only be passed one value each time: namely, FETCH after 20 ns, DECODE after 40 ns, and EXECUTE after 60 ns.

Chapter 10
VHDL in Use

This chapter will provide a set of four design examples which exploit various features of VHDL. They are chosen from different domains to emphasize the generality of the language. For those designers who are just becoming familiar with the language, these examples can provide a good starting point for learning VHDL. Each example is independent (*i.e.* does not depend on another example) so the reader is invited to start with an example from a familiar domain. The examples are shown in an order which best suits their presentation although in a real world situation it is unlikely this would be the order in which they are developed.

The first example is a simple device controller. The VHDL model makes use of some more sophisticated constructs of the language, namely, guarded signal assignments and resolved signals, to control the operation of three devices which share an output wire (and therefore must produce results at different times). This type of processing is often found in models at the system level.

The next example is a model of a T Flip Flop to which code has been added to verify the setup and hold restrictions of the device. It takes advantage of signal-valued attributes in the verification code and reports any violations using VHDL assertion statements. The example shows the flexible timing mechanisms of VHDL and the use of passive processes within entities. A VHDL description like this greatly reduces the amount of time to isolate errant behavior in a logic circuit. When the circuit does not operate correctly (or, more precisely, the VHDL simulation does not produce the expected results) the designer has a clear indication (*i.e.* an assertion violation) of where the error might lie.

The third example is the model of an element for constructing neural networks. Neural networks are interesting because the state of the network can be self-altering and, therefore, can be "taught" to perform certain classes of operations. The model presented here, though somewhat naive, is a good example of how VHDL can be used for more abstract modeling applications. A generic element for constructing neural networks is created and then used to instantiate three different networks performing the logical functions AND, OR, and XOR, respectively.

The final example is a behavioral model of a matrix multiplier implemented using systolic array processing. The circuit performs multiplication of any 3x3 matrix by a vector of length 3. Processing of this type can be found in the design of certain VLSI (and other) devices because the systolic algorithm reduces the number of processing elements needed to perform the multiplication. The VHDL model makes use of generics and generate statements to instantiate the required number of processing elements. The use of these constructs facilitates extending the model to handle differently sized matrices. This, therefore, represents an example of how to develop a modular, reusable design.

A Device Controller

The example presented in this section is a device controller. The design is not complete because the actual devices do not perform any real behavior, but the model is representative of a class of circuits which arise often in design situations. This example makes use of guarded signal assignments and resolved signals.

Before starting the description of the device controller, a package of functions for tri-state logic is developed. This package is illustrative of the type of package that might be defined for a team-wide project or collection of projects. The tristate type is defined and a number of associated types, functions, and resolution functions are defined for that type.

```
package TriState_Package is

    type TriState is ('X', '0', '1', 'Z');
    type TriState_Vector is array
    (Integer range <>) of TriState;
```

```
  function OrResolution (
    Inputs : TriState_Vector)
  return TriState;

  function AndResolution(
    Inputs : TriState_Vector)
  return TriState;

  function OneOnlyResolution(
    Inputs : TriState_Vector)
  return TriState;

  subtype WiredOr is OrResolution TriState;
  subtype WiredAnd is AndResolution TriState;
  subtype OneOnly is OneOnlyResolution TriState;

end TriState_Package;

package body TriState_Package is

  function OrResolution (
    Inputs : TriState_Vector)
  return TriState is
    variable Result : TriState := '0';
  begin
    for I in Inputs'Range loop
      if Inputs(I) = '1' then
        Result := '1';
      elsif Inputs(I) = 'X' then
        return 'X';
      end if;
    end loop;
    return Result;
  end OrResolution;

  function AndResolution (
    Inputs : TriState_Vector)
  return TriState is
    variable Result : TriState := '1';
  begin
    for I in Inputs'Range loop
      if Inputs(I) = '0' then
        Result := '0';
      elsif Inputs(I) = 'X' then
        return 'X';
      end if;
    end loop;
```

```
      return Result;
   end AndResolution;

   function OneOnlyResolution(
      Inputs : TriState_Vector)
   return TriState is
   begin
      if Inputs'Length /= 1 then
        if Inputs(Inputs'Left) = 'X' then
          return 'X';
        else
          assert (Inputs'Length >= 1)
            report "Under driven bus"
              severity Error;
          assert (Inputs'Length <= 1)
            report "Over driven bus"
              severity Error;
        end if;
      end if;
      return Inputs(Inputs'Low);
   end OneOnlyResolution;
end TriState_Package;
```

In order to present the device controller it is necessary to have some devices to control. The following entity defines the interface for the class of devices used in the example. This is followed by three architectural bodies which represent the device behavior. In this example the devices simply assert their activation.

```
   use work.TriState_Package.all;
   entity Device is
     port (
       Enable: in TriState;
       Result : out TriState);
   end Device;

   architecture Behavior1 of Device is
   begin
     process
     begin
       wait on Enable;
       if Enable = '1' then
         assert (False)
           report "Device 1 activated"
             severity Note;
```

```
            Result <= transport '0';
          end if;
        end process;
      end Behavior1;

      architecture Behavior2 of Device is
      begin
        process
        begin
          wait on Enable;
          if Enable = '1' then
            assert (False)
              report "Device 2 activated"
              severity Note;
            Result <= transport '1';
          end if;
        end process;
      end Behavior2;

      architecture Behavior3 of Device is
      begin
        process
        begin
          wait on Enable;
          if Enable = '1' then
            assert (False)
              report "Device 3 activated"
              severity Note;
            Result <= transport 'Z';
          end if;
        end process;
      end Behavior3;
```

An entity declaration is now defined for the device controller. The interface will have two ports; an input port and an output port. The input port is a vector of length two of tristate signals. This input will control the selection of one of the three devices: "01" will select the first device, "10" will select the second device, and "11" will select the third device. The output port is a single tristate signal which passes the result of an active device along the output line.

```
      use work.TriState_Package.all;
      entity Device_Network is
        port (
          C : TriState_Vector(1 to 2);
```

    Output : **out** TriState);
end Device_Network;

    Now the VHDL description of the controller is presented. The component for the device is declared and given the same interface as the entity declaration of the devices. Each device is represented inside a block statement. Each block is guarded by the value of the control signal C. Each block instantiates a different architecture for the device. The first concurrent statement in each block is a guarded signal assignment which enables the associated device. The guard on this assignment means that the device will only be enabled when the guard condition on the block becomes true, otherwise no assignment takes place. The Result signal is a resolved signal which takes the value of the device which is driving its result signal.

```
use work.TriState_Package.all;
architecture behavior of device_network is

   component SimpleDevice
      port (Enable: in TriState; Result : out TriState);
   end component;

   signal Result1, Result2, Result3: TriState := '0';
   signal Enable1, Enable2, Enable3: TriState := '0';
   signal Result: OneOnly register := 'X';

begin

   Device1_Block : block (C = ('0', '1'))
      for D1 : SimpleDevice use entity work.Device(Behavior1);
   begin
      Enable1 <= guarded '1';
      D1 : SimpleDevice port map (Enable1, Result1);
      Result <= guarded Result1;
   end block;

   Device2_Block : block (C = ('1', '0'))
      for D2 : SimpleDevice use entity work.Device(Behavior2);
   begin
      Enable2 <= guarded '1';
      D2 : SimpleDevice port map (Enable2, Result2);
      Result <= guarded Result2;
   end block;

   Device3_Block : block (C = ('1', '1'))
```

```
   for D3 : SimpleDevice use entity work.Device(Behavior3);
begin
   Enable3 <= guarded '1';
   D3 : SimpleDevice port map (Enable3, Result3);
   Result <= guarded Result3;
end block;

   Output <= Result;

end behavior;
```

Each block in this architecture has a guard condition which is dependent on the value of the signal C. When the guard becomes true, there is an assignment to the Enable signal for the device within the block. When the result comes out of the device it is assigned to the shared wire Result, but only if the guard condition is still true. Otherwise, the shared wire will not see the result of the device.

Finally, a test driver is developed to exercise the model and verify its operation. The test behavior cycles through the values of the control signal which enable each device.

```
entity test is end test;
use work.tristate_package.all;
architecture device of test is

   signal Input  : TriState_Vector(1 to 2) := ('0', '0');
   signal Output : TriState := 'X';

   component Device_Network
      port (C : TriState_Vector(1 to 2); Output : out TriState);
   end component;

   for UUT : Device_Network
      use entity work.Device_Network(Behavior);

begin

   UUT : Device_Network port map (Input, Output);

   Input <= transport
      ('0', '1') after 1 ns,
      ('1', '0') after 2 ns,
      ('1', '1') after 3 ns,
      ('0', '0') after 4 ns;
```

end device;

The results of simulating this model are given in the following table.

(NS)	Input(1 to 2)	Output
0	"00"	'X'
1	"01"	'0'
2	"10"	'1'
3	"11"	'Z'
4	"00"	

Setup and Hold Timing

An important part of verifying the timing of a model is checking that the model observes the constraints imposed by setup and hold times. Several features of VHDL offer the the designer considerable assistance in this task. Particularly useful are the assertion statement and the 'Stable attribute. This section gives an example of adding assertions to the description of a T flip-flop to check setup and hold timing constraints. The example also serves to show how timing can, in general, be made a function of generic parameters.

The following design entity provides a behavioral description of a T flip-flop. Note that the entity declaration declares generic parameters for setup and hold times.

```
entity TFF is
  generic
  (tPLH, tPHL : Time := 0 ns ;
   tPCL, tPPH : Time := 0 ns ;
   Setup, Hold : Time := 0 ns) ;
  port
  (T, CLK, PRESET, CLEAR : in Bit := '0' ;
   Q : inout Bit := '0' ;
   QB : inout Bit := '1' ) ;
end TFF ;

architecture BEHAVIOR of TFF is
begin

  Main:
  process (CLK, PRESET, CLEAR)
```

```
variable result : Bit ;
variable delay : Time := 0ns ;

begin

assert not (PRESET = '0' and CLEAR = '0')
  report "PRESET and CLEAR both '0' on T-Flip-Flop"
  severity Warning ;

if PRESET = '0' and CLEAR = '1'  then
  result := '1' ;
elsif PRESET = '1' and CLEAR = '0' then
  result := '0' ;
elsif CLK = '1' and not CLK'Quiet then
  if T = '1' then
    result := QB ;
  elsif T = '0' then
    result := Q ;
  end if ;
elsif CLK = '0' then
  result := Q ;
end if ;

if result = '0' then
  if CLEAR = '0' then delay := tPCL ;
  elsif Q = '1' then delay := tPHL ;
  end if ;
elsif result = '1' then
  if PRESET = '0' then delay := tPPH ;
  elsif Q = '0' then delay := tPLH ;
  end if ;
end if ;

Q <= result after delay ;
QB <= not result after delay ;

end process ;

end BEHAVIOR ;
```

Suppose that the designer wishes to check that the input signal T does not change within a minimum setup interval before the clock goes high and does not change within a minimum hold interval immediately after the clock goes high. To check this, he could add the following process statement to the entity declaration for TFF (since the process does not contain any signal assignments it is a passive process statement and may therefore occur in an entity declaration). The process statement is sensitive to changes on the signals CLK and T. If CLK has just gone high, then the time at which it has just gone high is saved in the variable Rising_at. If the value on T has just changed, then the time at which T has just changed is saved in the variable T_changed_at. The setup constraint is checked in the first of two assertions. If CLK has just gone high, then a warning is reported just in case the time interval between the present time (when CLK has just gone high) and the time when T last changed is not greater than some minimum setup time. The hold constraint is checked in the second assertion. If the value of T has just changed, then a warning is reported just in case the time interval between the present time (when the value of T has just changed) and the time when CLK last went high is not greater than some minimum hold time.

```
Check_setup_and_hold_times:
process (CLK, T)
  variable Rising_at : Time := 0 ns ;
  variable T_changed_at : Time := 0 ns ;
begin
  if not CLK'Stable and CLK = '1' then
    Rising_at := Now ;
  end if ;
  if not T'Stable then
    T_changed_at := Now ;
  end if ;
  if not CLK'Stable and CLK = '1' then
    assert Now - T_changed_at > Setup
      report "T changed within setup interval"
      severity Warning ;
  end if ;
  if not T'Stable then
    assert Now - Rising_at > Hold
      report "T changed within hold interval"
      severity Warning ;
  end if ;
end process ;
```

This check may be written in a simpler fashion by taking advantage of the fact that the 'Stable attribute can take a parameter specifying the length of an interval immediately preceding the present time. If an event occurred on T any time during the previous 2 ns, then the expression T'Stable (2 ns) will be false. Here is the simplified check:

```
      Check_setup_and_hold_times:
      process (CLK, T)
      begin
        if not CLK'Stable and CLK = '1' then
          assert T'Stable (Setup)
            report "T changed within setup interval"
            severity Warning ;
        end if ;
        if not T'Stable and CLK = '1' then
          assert CLK'Stable (Hold)
            report "T changed within hold interval"
            severity Warning ;
        end if ;
      end process ;
```

The entire T flip-flop description, together with a test-driver and
sample simulation output, follows.

```
    entity TFF is
      generic
      (tPLH, tPHL : Time := 0 ns ;
       tPCL, tPPH : Time := 0 ns ;
       Setup, Hold : Time := 0 ns) ;
      port
      (T, CLK, PRESET, CLEAR : in Bit := '0' ;
       Q : inout Bit := '0' ;
       QB : inout Bit := '1' ) ;
    begin

      Check_setup_and_hold_times:
      process (CLK, T)
      begin
        if not CLK'Stable and CLK = '1' then
          assert T'Stable (Setup)
            report "T changed within setup interval"
            severity Warning ;
        end if ;
        if not T'Stable and CLK = '1' then
          assert CLK'Stable (Hold)
            report "T changed within hold interval"
            severity Warning ;
        end if ;
      end process ;

    end TFF ;
```

```vhdl
architecture BEHAVIOR of TFF is
begin

  Main:
  process (CLK, PRESET, CLEAR)

    variable result : Bit ;
    variable delay : Time := 0ns ;

  begin

    assert not (PRESET = '0' and CLEAR = '0')
      report "PRESET and CLEAR both '0' on T-Flip-Flop"
      severity Warning ;

    if PRESET = '0' and CLEAR = '1'  then
      result := '1' ;
    elsif PRESET = '1' and CLEAR = '0' then
      result := '0' ;
    elsif CLK = '1' and not CLK'Quiet then
      if T = '1' then
        result := QB ;
      elsif T = '0' then
        result := Q ;
      end if ;
    elsif CLK = '0' then
      result := Q ;
    end if ;

    if result = '0' then
      if CLEAR = '0' then delay := tPCL ;
      elsif Q = '1' then delay := tPHL ;
      end if ;
    elsif result = '1' then
      if PRESET = '0' then delay := tPPH ;
      elsif Q = '0' then delay := tPLH ;
      end if ;
    end if ;

    Q <= result after delay ;
    QB <= not result after delay ;

  end process ;

end BEHAVIOR ;

entity Setup_tester is
```

```
end ;

architecture Setup_tester of Setup_tester is
  component  TFF
    generic (tPLH, tPHL : Time ;
             tPCL, tPPH : Time ;
             Setup, Hold : Time) ;
    port    (T, CLK, PRESET, CLEAR : in Bit ;
             Q, QB : inout Bit) ;
  end component ;
  for UUT : TFF use entity work.TFF (Behavior) ;
  signal T, CLK, PRESET, CLEAR, Q, QB : Bit ;
begin
  PRESET <= '1' ;
  CLEAR <= '1' ;
  T <=
    '1' after 5 ns ,
    '0' after 15 ns ,
    '1' after 25 ns ,
    '0' after 35 ns ,
    '1' after 45 ns - 4 ns ,
    '0' after 55 ns - 4 ns ,
    '1' after 65 ns + 4 ns ,
    '0' after 75 ns + 4 ns,
    '1' after 85 ns - 5 ns ,
    '0' after 95 ns - 5 ns ,
    '1' after 105 ns + 5 ns ,
    '0' after 115 ns + 5 ns ;

  CLK <= not CLK after 10 ns ;

  UUT : TFF
    generic map (10 ns, 10 ns, 10 ns, 10 ns, 2 ns, 2 ns)
    port map (T, CLK, PRESET, CLEAR, Q, QB) ;

  process (CLK)
  begin
    if Now >= 200 ns then
      assert False report "Stopping" severity Failure ;
    end if ;
  end process ;
end Setup_tester ;
```

The results of simulating this model are given in the following table.

(FS)	CLK	T	Q	QB
0	%% Assertion Violation (Warning) after 0 fs			
	PRESET and CLEAR both '0' on T-Flip-Flop			
	'0'	'0'	'0'	'1'
5000000		'1'		
10000000	'1'			
15000000		'0'		
20000000	'0'		'1'	'0'
25000000		'1'		
30000000	'1'			
35000000		'0'		
40000000	'0'		'0'	'1'
41000000		'1'		
50000000	'1'			
51000000	%% Assertion Violation (Warning) after 51 ns			
	T changed within hold interval			
		'0'		
60000000	'0'		'1'	'0'
69000000		'1'		
70000000	%% Assertion Violation (Warning) after 70 ns			
	T changed within setup interval			
		'1'		
79000000		'0'		
80000000	'0'	'1'	'0'	'1'
90000000	%% Assertion Violation (Warning) after 90 ns			
	T changed within setup interval			
	%% Assertion Violation (Warning) after 90 ns			
	T changed within hold interval			
	'1'	'0'		
100000000	'0'			
110000000	%% Assertion Violation (Warning) after 110 ns			
	T changed within setup interval			
	%% Assertion Violation (Warning) after 110 ns			
	T changed within hold interval			
	'1'	'1'		
120000000	'0'	'0'	'1'	'0'
130000000	'1'			
140000000	'0'			
150000000	'1'			
160000000	'0'			
170000000	'1'			
180000000	'0'			
190000000	'1'			
200000000	%% Assertion Violation (Fatal) after 200 ns			
	Stopping			

A Neural Net

One of the primary branches of the artificial intelligence field is computer learning. In recent years, researchers have made some headway with self-adjusting systems which exhibit some forms of learning. Many of these systems are based on the concept of neural networks; *i.e.* collections of nodes whose behavior is based upon a simplified model of the neurons which compose the human brain. It is beyond the scope of this book to examine the theory and operations of neural networks. The example contained here is a somewhat crude implementation of a McCulloch-Pitts perceptron which learns using back propagation. The model for this type of perceptron was originally developed by Frank Rosenblatt in the late fifties based on the work by Warren McCulloch and Walter Pitts in the late forties. Current work in this area is showing impressive results and the use of a language like VHDL for exchanging designs and implementations of neural networks could help researchers and developers to coordinate their efforts.

The example presented here is based on the following simple model of a neural processing element. Each element has a number of inputs and a single output. The output is computed by multiplying each input by its associated weight and adding them all together. If the sum is greater than or equal to the threshold value a 1 is applied to the output, otherwise the element generates a 0. This is depicted in Figure 10.1. The inputs, weights and threshold are represented by real numbers.

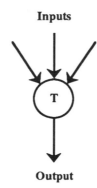

Figure 10.1. A Neural Network Element

The following package defines two types which will be used throughout the model of the neural network.

package Neural_Package **is**

```
type StimulusArray is array (Natural range <>) of Real;
type WeightArray is array (Natural range <>) of Real;

function CalculateSum (
  Values: StimulusArray;
  Weights : WeightArray)
  return Real;

end Neural_Package;
```

The first type in the package is an unconstrained array of real numbers. This type will be used to hold the collection of inputs. The second type is identical to the first type except in name; it is an unconstrained array of real numbers. This type will be used to hold the collection of weights which are applied to the input values. We distinguish between these two types in order to take advantage of the strong typing mechanisms in VHDL. This will guarantee that a collection of stimuli is never mistaken for a collection of weights or vice versa. However, since the elements of each array are real numbers it is possible to multiply a single element of a StimulusArray by a single element of a WeightArray. The package also defines a procedure called CalculateSum which takes an array of stimuli and their associated weights and computes the weighted sum. This function is defined in the package body.

```
package body Neural_Package is

  function CalculateSum (
    Values: StimulusArray;
    Weights : WeightArray)
    return Real
  is
    variable Sum : Real := 0.0;
  begin
    for I in Values'Range loop
      Sum := Sum + Values(I) * Weights(I);
    end loop;
    return Sum;
  end CalculateSum;

end Neural_Package;
```

The sum is initialized to 0 and a loop based on the number of inputs ('Range gives the actual index range of the input; in this case 1 to 2) is used to add in each product of a stimulus and its weight.

Each element used in the neural network is identical except for three variants: the collection of inputs, the collection of weights for those inputs, and the threshold value of the element. The only structural variants, therefore, are the number of inputs (and weights); all other variation is value driven. Therefore, the neural processing element must be designed to accommodate any number of inputs. Fortunately this is quite simple in VHDL since a port may be an unconstrained array and the attributes of the port can be used to determine the actual size within the architecture. Based on this information, the following entity declaration is used.

```
use work.Neural_Package.all;
entity Neural_Element is
  generic (
    Threshold : Real);
  port (
    Stimulus: in StimulusArray;
    Weights : in WeightArray;
    Output: out Real := 0.0);
  end Neural_Element;
```

The generic determines the threshold of the element. The value for the threshold is passed in from an instantiating unit. Two inputs are passed in from the instantiating component: the array of stimulus signals and the array of weights for each of the stimulus signals. Finally, an output is defined to pass out the result of the network's computation.

The behavior of the neural processing element can be summed up in the following statements.

1.   Compute the sum by adding the result of each input multiplied by its weight.

2.   If the sum is greater than or equal to the threshold value of the element assign a 1 to the output of the element, otherwise assign 0.

This behavior is straightforward to describe in VHDL. The process statement can be viewed in two sections, each corresponding to one of the statements above. The behavior of the neural processing element is expressed in the following architecture:

```
architecture behavior of neural_element is
begin
  NeuralProcess:
  process
    variable Sum : Real;
```

```
begin
  wait on Stimulus;
  Sum := CalculateSum(Stimulus, Weights);
  if Sum >= Threshold then
    Output <= 1.0;
  else
    Output <= 0.0;
  end if;
 end process;
end behavior;
```

A few things should be noted about the process statement in this architecture. First, the variable declaration defines Sum to be visible only within this process. It is given no initial value because it will be initialized each time before it is used. The process statement begins with a wait statement. Remember that each process in a model executes at the start of simulation. In the case of this element, the processing should only occur when stimulus is applied to the inputs, so the initial execution of the process will just cause the process to suspend until activity occurs on the inputs. The sum is then determined by calling the function CalculateSum with the stimulus and weights. The next section chooses an assignment to the output based on the comparison of the sum to the threshold value.

The processing element is now completed. The next step is to test the element with a few sample cases. The networks which are instantiated are the logic functions OR, AND, and XOR. The networks for these functions are depicted in Figure 10.2.

The VHDL test bench for these circuits is given by the following architecture.

```
entity test is end test;
use work.neural_package.all;
architecture neural of test is
  -- Stimulus Signal for All Networks
  signal Stimulus : StimulusArray(1 to 2) := (0.0, 0.0);

  -- Constants and Signals for OR  network
  signal OrOutput : Real := 0.0;
  signal OrWeights : WeightArray(1 to 2) := (1.0, 1.0);

  -- Constants and Signals for AND network
  signal AndOutput : Real := 0.0;
  signal AndWeights : WeightArray(1 to 2) := (0.5, 0.5);

  -- Constants and Signals for XOR network
```

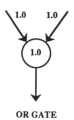

OR GATE

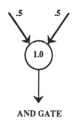

AND GATE

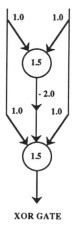

XOR GATE

Figure 10.2. Neural Networks for OR, AND, and XOR Functions

```
signal XorOutput : Real := 0.0;
signal XorWeights1 : WeightArray(1 to 2) := (1.0, 1.0);
signal XorWeights2 : WeightArray(1 to 3) := (1.0, -2.0, 1.0);
signal HiddenStimulus : StimulusArray(1 to 3)
    := (0.0, 0.0, 0.0);

-- Component Declaration for a Neural Element
component Neural_Node
  generic (
    Threshold : Real);
  port (
    Stimulus: StimulusArray;
    Weights : WeightArray;
    Output: out Real);
end component;

-- Instantiation of all Neural_Nodes to
--   Neural_Element design unit
for all : Neural_Node use
    entity work.Neural_Element(Behavior);
```

begin
   -- Instantiation of Node Elements for OR Network
   OrElement1 : Neural_Node
     generic map (1.0)
     port map (Stimulus, OrWeights, OrOutput);

   -- Instantiation of Node Elements for AND Network
   AndElement1 : Neural_Node
     generic map (1.0)
     port map (Stimulus, AndWeights, AndOutput);

   -- Instantiation of Node Elements for XOR Network
   XorElement1 : Neural_Node
     generic map (1.5)
     port map (Stimulus, XorWeights1, HiddenStimulus(2));
   XorElement2 : Neural_Node
     generic map (0.5)
     port map (HiddenStimulus, XorWeights2, XorOutput);

   HiddenStimulus(1) <= Stimulus(1);
   HiddenStimulus(3) <= Stimulus(2);

   Stimulus <= **transport**
     (0.0, 0.0) **after** 0 ns,
     (0.0, 1.0) **after** 1 ns,
     (1.0, 0.0) **after** 2 ns,
     (1.0, 1.0) **after** 3 ns;

end Neural;

This test bench tests three instantiations of networks. The constants and signals for each test are grouped together under VHDL comments. The stimulus for each network is the same, namely, all possible states for a two input logic function. In the XOR case, two elements are used because the XOR function requires a "hidden" element in the network.

    The results of simulating this model are given in the following table.

(NS)	Stimulus(1)	Stimulus(2)	OrOutput	AndOutput	XorOutput
0	0.0	0.0	0.0	0.0	0.0
1	0.0	1.0	1.0	0.0	1.0
2	1.0	0.0	1.0	0.0	1.0
3	1.0	1.0	1.0	1.0	0.0

A Systolic Array Multiplier

A systolic array is a collection of processors arranged in an array. Data is passed from a given processor to neighboring processors in regular patterns. Data is passed through the array and into the array in "beats" (*i.e.* regular intervals) which explains the use of the term *systolic* (like a heartbeat).

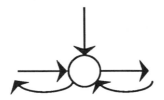

Figure 10.3. Processing Element of a Systolic Array

In the example presented here, each processor has the form depicted in Figure 10.3. Each processor has three inputs: one each for data passed in from the left of the processor, the top of the processor and the right of the processor. Each processor passes data out to its left and to its right. In many applications, including this one, these processors are laid out in regular patterns in order to perform operations. The example which is given here is the multiplication of a matrix by an vector, starting with a 3x3 matrix by a vector of length 3. The regular array of processors needed for this application is shown in Figure 10.4.

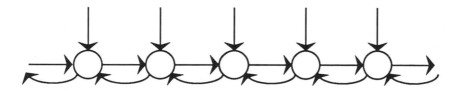

Figure 10.4. Systolic Array for Vector Multiplication of a 3x3 Matrix

The multiplication of a matrix by a vector is a straightforward operation given by the equation:

$$c_k = \sum_{i=1}^{i=n} A_{ik} b_i$$

where A is the matrix, b is the vector and c is the resulting vector. By "beating in" the vector from the left and the matrix from the top, the result "beats" out of the output on the left. This process is depicted in Figure 10.5. There is a beat between each input of b and each pass of c to the rightmost element.

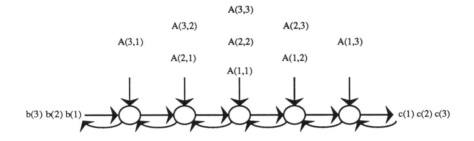

Figure 10.5. Phased Input for Vector Multiplication

VHDL does not include a predefined matrix type so one must be created. The following package defines a type Matrix which is unconstrained for both rows and columns and an unconstrained type Vector.

```
package matrix_package is

    subtype Dimension is Natural;
    type Vector is array
        (Dimension range <>) of Integer;
    type Matrix is array
        (Dimension range <>, Dimension range <>) of Integer;

end matrix_package;
```

As stated above, each element of a systolic array has three inputs and two outputs. The interface for the systolic array element should therefore have three input ports and two output ports. The significant feature of each of these ports is its position around the array, so their names will be chosen to reflect their position. The following entity declaration defines the interface to an element of a systolic array.

```
entity systolic_element is
    port (
        Left_In, Top_In, Right_In : in Integer := 0;
```

Left_Out, Right_Out : **out** Integer := 0);
end systolic_element;

The algorithm for matrix multiplication using a systolic array requires that each element perform the following two operations:

1. Pass through the input on the left to the output on the right.
2. Multiply the left input by the top input, add the right input to this product and pass the result out through the left.

The following architecture implements this algorithm directly.

```
architecture matrix_multiplication of systolic_element is
begin
  process
    (Left_In'Transaction,
     Top_In'Transaction,
     Right_In'Transaction)
    variable FirstTime : Boolean := True;
  begin
    Right_Out <= Left_In after 1 ns;
    Left_Out <= (Left_In * Top_In) + Right_In after 1 ns;
  end process;
end matrix_multiplication;
```

Now that the processing element is established it is possible to create systolic arrays which make use of these processors. The first application presented is a systolic array processor which performs multiplication of a 3x3 matrix by a vector of length 3. The following interface defines a design entity which takes two inputs, a 3x3 matrix (A) and a vector of length 3 (b) and produces an output vector of length 3 (c).

```
use work.matrix_package.all;
entity systolic_processor is
  port (
    A : in Matrix(1 to 3, 1 to 3);
    b : in Vector(1 to 3);
    c : out Vector(1 to 3));
end systolic_processor;
```

The body of the systolic array processor is divided into three functional sections. The first section generates the elements of the array using array signals to hold the intermediate values being passed through the array. The second section "beats" in the values to the inputs of the processors. The final section catches the values being passed out of the leftmost element of the array and places them in the result vector.

```
architecture behavior of systolic_processor is

   component systolic_element
     port (
       Left_In : in Integer;
       Top_In : in Integer;
       Right_In: in Integer;
       Left_Out: out Integer;
       Right_Out : out Integer);
   end component;

   type Systolic_Array is array (integer range <>) of Integer;
   signal TI : Systolic_Array(1 to 5) := (0, 0, 0, 0, 0);
   signal LO : Systolic_Array(1 to 5) := (0, 0, 0, 0, 0);
   signal RO : Systolic_Array(1 to 5) := (0, 0, 0, 0, 0);

   signal Zero : Integer := 0;
   signal FeedIn : Integer := 0;

   for all : systolic_element use
     entity work.systolic_element(matrix_multiplication);

begin

   -- Generate the elements of the array
   Element_Generation:
   for I in 1 to 5 generate

     First_Element:
     if I = 1 generate
       L1 : systolic_element
         port map (FeedIn, TI(I), LO(I+1), LO(I), RO(I));
     end generate;

     Middle_Elements:
     if I > 1 and I < 5 generate
       L2 : systolic_element
         port map (RO(I-1), TI(I), LO(I+1), LO(I), RO(I));
     end generate;

     Last_Element:
     if I = 5 generate
```

```
        L5 : systolic_element
            port map (RO(I-1), TI(I), Zero, LO(I), RO(I));
        end generate;

    end generate;

    -- Beat in A into the top of the array
    TI <=
        (0, 0, A(1,1), 0, 0) after 3 ns,
        (0, A(2,1), 0, A(1,2), 0) after 4 ns,
        (A(3,1), 0, A(2,2), 0, A(1,3)) after 5 ns,
        (0, A(3,2), 0, A(2,3), 0) after 6 ns,
        (0, 0, A(3,3), 0, 0) after 7 ns;

    -- Beat in the b vector into the leftmost input of array
    FeedIn <=
        b(1) after 1 ns,
        0 after 2 ns,
        b(2) after 3 ns,
        0 after 4 ns,
        b(3) after 5 ns,
        0 after 6 ns;

    -- Read c from the leftmost output of the array
    process (LO(1)'Transaction)
        variable index : integer := 1;
        variable count : integer := -1;
    begin
        assert (index < 4) report "Completed" severity Error;
        count := count + 1;
        if count = 6 or count = 8 or count = 10 then
            c(index) <= LO(1);
            index := index + 1;
        end if;
    end process;

    end Behavior;
```

The declarative region of the architecture defines the component declaration that is needed for instantiating each element of the systolic array. Then a type (Systolic_Array) is defined which is then used to create the signal arrays for the intermediate values being passed through the array. A signal Zero is defined which always has the value 0. This signal is used as the rightmost input of the element array.

The body of the architecture begins by generating instantiations of all the systolic elements needed for the processing. Five elements are generated with the first and last element being treated specially because the first element will get its values from the b vector and the last element takes an input of zero. The next section of the architecture pulses in the values of the matrix into the top of the systolic array and the values of the vector into FeedIn. Finally, the values coming out of the leftmost element of the array are read into the vector c.

The test bench for this design is straightforward. Two multiplications are chosen to verify that the systolic implementation performs the correct operations.

```
entity test is end test;
use work.matrix_package.all;
architecture systolic of test is

  signal A : Matrix(1 to 3, 1 to 3) := ((0,0,0),(0,0,0),(0,0,0));
  signal b, c : Vector(1 to 3);

  component Systolic_Processor
    port (
      A : Matrix(1 to 3, 1 to 3);
      b : Vector(1 to 3);
      c : out Vector(1 to 3));
  end component;

  for L1 : Systolic_Processor use
    entity work.systolic_processor(behavior);

begin

  L1 : Systolic_Processor
    port map (A, b, c);

  A <= transport ((3, 0, 0), (0, 2, 0), (0, 0, 1));
  b <= transport (2, 2, 2);

end systolic;
```

Finally, a configuration body is used to bind the components to the design entities in the library. This configuration does not include explicit binding indications so the components pick up entities whose names match those of the components.

```
use work.all;
configuration conf of test is
  use work.all;
  for systolic
  end for;
end conf;
```

Summary

The examples presented in this chapter are not meant to cover the full expressive range of VHDL. They are included here to give the reader an idea of the various ways in which VHDL can be applied to real world problems. As VHDL evolves, and as the users of VHDL become more fluent in VHDL, the range of applications will be extended. These examples provide a good starting point for beginning to use VHDL, but the real learning tool is experience.

Appendix A
Predefined Environment

VHDL contains a number of attributes and packages which are predefined and must be included by all implementations of VHDL.

Reserved Words

The following identifiers are reserved in VHDL, meaning they cannot be used as identifiers in a VHDL model.[1]

abs	else	map	select
access	elsif	mod	severity
after	end		signal
alias	entity	nand	subtype
all	exit	new	
and		next	then
architecture		nor	to

---

[1] This table of reserved words is reprinted from IEEE Std 1076-1987 *IEEE Standard VHDL Language Reference Manual*, Copyright © 1988 by the Institute of Electrical and Electronics Engineers, Inc., and is reproduced here by permission of the IEEE Standards Department.

array	file	not	transport
assert	for	null	type
attribute	function		
		of	units
begin	generate	on	until
block	generic	open	use
body	guarded	or	
buffer		others	variable
bus	if	out	
	in		wait
case	inout	package	when
component	is	port	while
configuration		procedure	with
constant		process	
	label		xor
disconnect	library	range	
downto	linkage	record	
	loop	register	
		rem	
		report	
		return	

Attributes

Type and Subtype Attributes

The following attributes are defined for types and subtypes:

T'BASE

> This attribute returns the base type of type T. It is only legal when used as the prefix to another attribute; e.g. T'BASE'RIGHT.

T'LEFT

> This attribute returns the left bound of scalar type T.

T'RIGHT

> This attribute returns the right bound of scalar type T.

T'HIGH

> This attribute returns the upper bound of scalar type T.

T'LOW

> This attribute returns the lower bound of scalar type T.

T'POS(X)

> This attribute returns the position within the discrete or physical type of the value of the parameter.

T'VAL(X)

> This attribute returns the value at the position within the discrete or physical type which is given by X.

T'SUCC(X)

> This attribute returns the value at the position within the discrete or physical type whose position is one greater than the parameter.

T'PRED(X)

> This attribute returns the value at the position within the discrete or physical type whose position is one less than the parameter.

T'LEFTOF(X)

> This attribute returns the value at the position within the discrete or physical type whose position is to the left of the parameter.

T'RIGHTOF(X)

> This attribute returns the value at the position within the discrete or physical type whose position is to the right of the parameter.

Array Attributes

The following attributes are defined for array objects or constrained array subtypes.

A'LEFT(N)

The parameter is optional; the default is 1. This attribute returns the left bound of the of the Nth index of the array object or subtype.

A'RIGHT(N)

The parameter is optional; the default is 1. This attribute returns the right bound of the of the Nth index of the array object or subtype.

A'HIGH(N)

The parameter is optional; the default is 1. This attribute returns the upper bound of the of the Nth index of the array object or subtype.

A'LOW(N)

The parameter is optional; the default is 1. This attribute returns the lower bound of the of the Nth index of the array object or subtype.

A'RANGE(N)

The parameter is optional; the default is 1. This attribute returns the range of the Nth index of the array object or constrained array subtype. If the object or subtype is ascending, the range *left bound* **to** *right bound* is returned; otherwise, the range *left bound* **downto** *right bound* is returned.

A'REVERSE_RANGE(N)

The parameter is optional; the default is 1. This attribute is identical to A'RANGE(N) *except* that the range is reverse; *i.e.* for ascending, *right bound* **downto** *left bound* and for descending, *right bound* **to** *left bound*.

A'LENGTH(N)

The parameter is optional; the default is 1. This attribute returns the number of values in the Nth index of the array object or constrained array subtype.

Signal-Valued Attributes

The following attributes are signal-valued; *i.e.* each occurrence of one of these attributes is a signal which may, for instance, be placed in the sensitivity list of a process.

S'DELAYED(T)

The parameter is optional; the default is 0 ns. This attribute defines a signal whose value is the value of S delayed by the time T. If T is 0 ns, the value is equal to S after a delta delay (*i.e.* in the next simulation cycle).

S'STABLE(T)

The parameter is·optional; the default is 0 ns. This attribute defines a boolean signal whose value is TRUE if S has not had an event (*i.e.* not changed value) for the length of time T; otherwise, the value of the signal is FALSE. If T is 0 ns, then S will be FALSE during the simulation cycle in which S changed and then will return to TRUE.

S'QUIET(T)

The parameter is optional; the default is 0 ns. This attribute defines a boolean signal whose is TRUE if S has not had a transaction (*i.e.* not active) for the length of time T; otherwise, the value of the signal is FALSE. If T is 0 ns, then S will be FALSE during the simulation cycle in which S was assigned to and then will return to TRUE.

S'TRANSACTION

This attribute defines a bit signal whose value toggles each time a transaction occurs on S (*i.e.* S is active).

Signal-Related Attributes

The following attributes are defined for signal objects but are not signals themselves:

S'EVENT

This boolean attribute is true if an event has occurred on S during the current simulation cycle (*i.e.* if S has changed value during the cycle).

S'ACTIVE

This boolean attribute is true if a transaction has occurred on S during the current simulation cycle (*i.e.* if S is active during the cycle).

S'LAST_EVENT

> This time-valued attribute returns the amount of time which has elapsed since the last event on S (*i.e.* since S last changed value).

S'LAST_ACTIVE

> This time-valued attribute returns the amount of time which has elapsed since the last transaction on S (*i.e.* since S was last active).

S'LAST_VALUE

> This attribute returns the value of S before the last event on S.

> The following attributes are defined for blocks and design entities.

B'BEHAVIOR

> This attribute is true if there are no component instantiation statements in the block or architecture.

B'STRUCTURE

> This attribute is true if, within the block or architecture, all process statements or equivalent process statements are passive (*i.e.* do not contain signal assignments).

Packages

The Package STANDARD

The package STANDARD is implicitly used by all design entities. Its source is defined below.[2]

package STANDARD **is**

-- predefined enumeration types:

---

[2] This VHDL package is reprinted from IEEE Std 1076-1987 *IEEE Standard VHDL Language Reference Manual*, Copyright © 1988 by the Institute of Electrical and Electronics Engineers, Inc., and is reproduced here by permission of the IEEE Standards Department.

type BOOLEAN **is** (FALSE, TRUE) ;

type BIT **is** ('0', '1') ;

type CHARACTER **is** (
NUL, SOH, STX, ETX, EOT, ENQ, ACK, BEL,
BS, HT, LF, VT, FF, CR, SO, SI,
DLE, DC1, DC2, DC3, DC4, NAK, SYN, ETB,
CAN, EM, SUB, ESC, FSP, GSP, RSP, USP,

' ', '!', '"', '#', '$', '%', '&', ''',
'(', ')', '*', '+', ',', '-', '.', '/',
'0', '1', '2', '3', '4', '5', '6', '7',
'8', '9', ':', ';', '<', '=', '>', '?',

'@', 'A', 'B', 'C', 'D', 'E', 'F', 'G',
'H', 'I', 'J', 'K', 'L', 'M', 'N', 'O',
'P', 'Q', 'R', 'S', 'T', 'U', 'V', 'W',
'X', 'Y', 'Z', '[', '\', ']', '^', '_',

'`', 'a', 'b', 'c', 'd', 'e', 'f', 'g',
'h', 'i', 'j', 'k', 'l', 'm', 'n', 'o',
'p', 'q', 'r', 's', 't', 'u', 'v', 'w',
'x', 'y', 'z', '{', '|', '}', '~', DEL) ;

type SEVERITY_LEVEL **is**
(NOTE, WARNING, ERROR, FAILURE) ;

-- predefined numeric types:

type INTEGER **is range** *implementation_defined*;
type REAL **is range** *implementation_defined*;

-- predefined type TIME:

type TIME **is range** *implementation_defined*
  units
    fs ; -- femtosecond
    ps = 1000 fs ; -- picosecond
    ns = 1000 ps ; -- nanosecond
    us = 1000 ns ; -- microsecond
    ms = 1000 us ; -- millisecond
    sec = 1000 ms ; -- second
    min = 60 sec ; -- minute
    hr = 60 min ; -- hour

end units ;

-- function that returns the current simulation time:

function NOW **return** TIME;

-- predefined numeric subtypes:

subtype NATURAL **is**
    INTEGER **range** 0 **to** INTEGER'HIGH;

subtype POSITIVE **is**
    INTEGER **range** 1 **to** INTEGER'HIGH;

-- predefined array types:

type STRING **is**
    array (POSITIVE **range** <>) **of** CHARACTER;

type BIT_VECTOR **is**
    array (NATURAL **range** <>) **of** BIT;

end STANDARD;

The Package TEXTIO

Package TEXTIO defines a number of routines which are used to read and write ASCII files (and terminals). The source for TEXTIO is given below.[3]

package TEXTIO **is**

-- Type definitions for Text I/O

type LINE **is access** STRING;

---

[3] This VHDL package is reprinted from IEEE Std 1076-1987 *IEEE Standard VHDL Language Reference Manual*, Copyright © 1988 by the Institute of Electrical and Electronics Engineers, Inc., and is reproduced here by permission of the IEEE Standards Department.

-- A LINE is a pointer to a STRING value

type TEXT **is file of** STRING;
  -- a file of variable-length ASCII records

type SIDE **is** (RIGHT, LEFT);
  -- for justifying output data within fields

subtype WIDTH **is** NATURAL;
  -- for specifying widths of output fields

-- Standard Text Files

file INPUT : TEXT **is in** "STD_INPUT";
file OUTPUT : TEXT **is out** "STD_OUTPUT";

-- Input Routines for Standard Types

procedure READLINE (F: **in** TEXT; L: **out** LINE);

procedure READ
  (L: **inout** LINE; VALUE: **out** BIT;
   GOOD: **out** BOOLEAN);
procedure READ
  (L: **inout** LINE; VALUE: **out** BIT);

procedure READ
  (L: **inout** LINE; VALUE: **out** BIT_VECTOR;
   GOOD: **out** BOOLEAN);
procedure READ
  (L: **inout** LINE; VALUE: **out** BIT_VECTOR);

procedure READ
  (L: **inout** LINE; VALUE: **out** BOOLEAN;
   GOOD: **out** BOOLEAN);
procedure READ
  (L: **inout** LINE; VALUE: **out** BOOLEAN);

procedure READ
  (L: **inout** LINE; VALUE: **out** CHARACTER;
   GOOD: **out** BOOLEAN);
procedure READ
  (L: **inout** LINE; VALUE: **out** CHARACTER);

procedure READ

```
    (L: inout LINE; VALUE: out INTEGER;
    GOOD: out BOOLEAN);
procedure READ
    (L: inout LINE; VALUE: out INTEGER);

procedure READ
    (L: inout LINE; VALUE: out REAL;
    GOOD: out BOOLEAN);
procedure READ
    (L: inout LINE; VALUE: out REAL);

procedure READ
    (L: inout LINE; VALUE: out STRING;
    GOOD: out BOOLEAN);
procedure READ
    (L: inout LINE; VALUE: out STRING);

procedure READ
    (L: inout LINE; VALUE: out TIME;
    GOOD: out BOOLEAN);
procedure READ
    (L: inout LINE; VALUE: out TIME);

-- Output Routines for Standard Types

procedure WRITELINE(F: out TEXT; L: in LINE);

procedure WRITE
    (L: inout LINE; VALUE: in BIT;
    JUSTIFIED: in SIDE:= RIGHT; FIELD:in WIDTH := 0);

procedure WRITE
    (L: inout LINE; VALUE: in BIT_VECTOR;
    JUSTIFIED: in SIDE:= RIGHT; FIELD:in WIDTH := 0);

procedure WRITE
    (L: inout LINE; VALUE: in BOOLEAN;
    JUSTIFIED: in SIDE:= RIGHT; FIELD:in WIDTH := 0);

procedure WRITE
    (L: inout LINE; VALUE: in CHARACTER;
    JUSTIFIED: in SIDE:= RIGHT; FIELD:in WIDTH := 0);

procedure WRITE
    (L: inout LINE; VALUE: in INTEGER;
    JUSTIFIED: in SIDE:= RIGHT; FIELD:in WIDTH := 0);
```

```
procedure WRITE
  (L: inout LINE; VALUE: in REAL;
   JUSTIFIED: in SIDE:= RIGHT; FIELD:in WIDTH := 0;
   DIGITS: in NATURAL := 0);

procedure WRITE
  (L: inout LINE; VALUE: in STRING;
   JUSTIFIED: in SIDE:= RIGHT; FIELD:in WIDTH := 0);

procedure WRITE
  (L: inout LINE; VALUE: in TIME;
   JUSTIFIED: in SIDE:= RIGHT; FIELD:in WIDTH := 0;
   UNIT: in TIME := NS);

-- File Position Predicates

function ENDLINE(L: in LINE) return BOOLEAN;
-- function ENDFILE(F: in TEXT) return BOOLEAN;

end TEXTIO;
```

Appendix B
VHDL Syntax

The syntax in this appendix is derived from the syntax contained in Appendix A of the IEEE Standard VHDL Language Reference Manual. While the syntax here presented is considerably simpler than that contained in the standard reference manual, it nevertheless describes the same language.

Each of the following syntax rules consists of two parts, a left-hand part and a right-hand part. The two parts are separated by the single arrow symbol (→). The left-hand part is the name of a syntactic construct, and the right-hand part describes the internal structure of the construct. If there are several ways of forming the construct named in the left-hand, then the right-hand will contain alternative ways of forming the construct, the alternatives being separated by the sharp symbol (#). Any sequence of symbols enclosed in square brackets ([and]) is optional; that is, there is an alternative which contains the sequence and another alternative which omits the sequence. Any sequence of symbols enclosed in curly brackets ({ and }) may occur zero or more times; that is, there is an unlimited number of alternatives: one which omits the sequence, one which has one occurrence of the sequence, one which has two occurrences of the sequence, etc.

Three typefaces are used in the syntax rules: roman, bold, and italic. All grammar symbols that can appear in the left-hand part of a rule are written in lowercase roman. Reserved words are always atomic;

that is, a reserved word is not itself a construct and therefore does not occur in the left-hand of any syntax rule. The names of reserved words appear in lowercase bold, as elsewhere in this book. There are a handful of other atomic symbols, whose internal structure is not given in these rules; these symbols are written in uppercase roman (IDENTIFIER, CHARACTER-LITERAL, BIT-STRING-LITERAL, STRING-LITERAL, INTEGER-LITERAL, FLOATING-POINT-LITERAL). Some symbols consist of a lowercase italic prefix and a lowercase roman main part. The italic prefix has no syntactic significance and is provided only to elucidate the semantic function of the symbol. Two symbols having the same main part but differing in their italic prefixes are syntactically identical. There will be a rule decomposing the main part of such a symbol, but there will be no rule whose left-hand contains an italic prefix.

The syntax rules are arranged in groups of related rules. The groups are:

> Literals and Miscellaneous
> Design Unit
> Library Units
> Declarative Item
> Subprograms
> Interface Lists and Association Lists
> Names and Expressions
> Operators
> Element Association and Choices
> Type Declarations
> Subtypes and Constraints
> Objects, Aliases, Files, Disconnections
> Attribute Declarations and Specifications
> Schemes
> Concurrent Statements
> Sequential Statements
> Components and Configurations

This grouping is followed by an alphabetical index of rules.

```
---------------------------------------------------
--          Literals and Miscellaneous
---------------------------------------------------
```

1. designator →
     IDENTIFIER
     # STRING-LITERAL

2. literal →
     abstract-literal
     # enumeration-literal
     # BIT-STRING-LITERAL
     # STRING-LITERAL
     # physical-literal
     # **null**

3. abstract-literal →
     INTEGER-LITERAL
     # FLOATING-POINT-LITERAL

4. enumeration-literal →
     CHARACTER-LITERAL
     # IDENTIFIER

5. physical-literal →
     [abstract-literal] IDENTIFIER

6. identifier-list →
     IDENTIFIER { , IDENTIFIER }

```
---------------------------------------------------
--                Design Unit
---------------------------------------------------
```

7. design-unit →
     { context-item }
     library-unit

8. library-unit →
     entity-declaration
     # configuration-declaration
     # package-declaration
     # architecture-body
     # package-body

9. context-item →
      library-clause
      # use-clause

10. library-clause →
      library identifier-list ;

11. use-clause →
      use selected-name { , selected-name } ;

```
-----------------------------------------------------
--              Library Units
-----------------------------------------------------
```

12. entity-declaration →
      entity IDENTIFIER **is**
      [**generic** interface-list ;]
      [**port** interface-list ;]
      [**begin**
      concurrent-statements]
      end [IDENTIFIER] ;

13. architecture-body →
      architecture IDENTIFIER **of** *entity*-mark **is**
        { declarative-item }
      begin
        concurrent-statements
      end [IDENTIFIER] ;

14. configuration-declaration →
      configuration IDENTIFIER **of** *entity*-mark **is**
        { declarative-item }
        block-configuration
      end [IDENTIFIER] ;

15. package-declaration →
      package IDENTIFIER **is**
        { declarative-item }
      end [IDENTIFIER] ;

16. package-body →
      package body IDENTIFIER **is**
        { declarative-item }
      end [IDENTIFIER] ;

```
----------------------------------------------------
--            Declarative Item
----------------------------------------------------
```

17. declarative-item →
    type-declaration
    # subtype-declaration
    # object-declaration
    # file-declaration
    # alias-declaration
    # subprogram-declaration
    # subprogram-body
    # component-declaration
    # attribute-declaration
    # attribute-specification
    # configuration-specification
    # disconnection-specification
    # use-clause

```
----------------------------------------------------
--            Subprograms
----------------------------------------------------
```

18. subprogram-declaration →
    subprogram-specification ;

19. subprogram-specification →
    procedure designator [interface-list]
    # **function** designator [interface-list] **return** *type*-mark

20. subprogram-body →
    subprogram-specification **is**
      { declarative-item }
    begin
      sequential-statements
    end [designator] ;

```
----------------------------------------------------
--   Interface Lists and Association Lists
----------------------------------------------------
```

21. interface-list →
    (interface-element { ; interface-element })

22. interface-element →
    [object-class] identifier-list : [mode] subtype-indication
    [**bus**] [:= expression]

23. mode →
    in
    # **out**
    # **inout**
    # **buffer**
    # **linkage**

24. association-list →
    (association-element { , association-element })

25. association-element →
    [formal-part =>] actual-part

26. formal-part →
    name
    # *function*-mark (name)

27. actual-part →
    expression
    # **open**

--   Names and Expressions
----

28. mark →
    IDENTIFIER
    # selected-name

29. expression →
    primary
    # unary-operator primary
    # primary { binary-operator primary }

30. primary →
    name
    # literal
    # aggregate
    # function-call
    # qualified-expression
    # type-conversion
    # allocator
    # (expression)

31. name →
    designator

        # selected-name
        # indexed-name
        # slice-name
        # attribute-name

32. selected-name →
        prefix . suffix

33. prefix →
        name
        # function-call

34. suffix →
        designator
        # CHARACTER-LITERAL
        # **all**

35. indexed-name →
        prefix (expression { , expression })

36. slice-name →
        prefix (discrete-range)

37. attribute-name →
        prefix ' *attribute*-IDENTIFIER [(*static*-expression)]

38. function-call →
        function-mark [association-list]

39. aggregate →
        (element-association { , element-association })

40. qualified-expression →
        type-mark ' (expression)
        # *type*- mark ' aggregate

41. type-conversion →
        type-mark (expression)

42. allocator →
        new subtype-indication
        # **new** qualified-expression

--   Operators
----

43. binary-operator →
       +
       # -
       # *
       # /
       # **
       # **mod**
       # **rem**
       # &
       # =
       # <
       # >
       # <=
       # >=
       # /=
       # **and**
       # **or**
       # **xor**
       # **nand**
       # **nor**

44. unary-operator →
       abs
       # **not**
       # +
       # -

--   Element Association and Choices
----

45. element-association →
       [choices =>] expression

46. choices →
       choice { | choice }

47. choice →
       expression
       # discrete-range
       # **others**

--   Type Declarations
----

48. type-declaration →

type IDENTIFIER [**is** type-definition] ;

49. type-definition →
    enumeration-type-definition
    # range-constraint
    # physical-type-definition
    # unconstrained-array-definition
    # constrained-array-definition
    # record-type-definition
    # access-type-definition
    # file-type-definition

50. enumeration-type-definition →
    (enumeration-literal { , enumeration-literal })

51. physical-type-definition →
    range-constraint
      units
        base-unit-declaration
        { secondary-unit-declaration }
      end units

52. base-unit-declaration →
    IDENTIFIER ;

53. secondary-unit-declaration →
    IDENTIFIER = physical-literal ;

54. unconstrained-array-definition →
    array (index-subtype-definition { , index-subtype-definition })
    of subtype-indication

55. index-subtype-definition →
    type-mark **range** <>

56. constrained-array-definition →
    array index-constraint **of** subtype-indication

57. record-type-definition →
    record
      element-declaration
      { element-declaration }
    end record

58. element-declaration →
    identifier-list : subtype-indication ;

59. access-type-definition →
    access subtype-indication

60. file-type-definition →
    file of *type*-mark

```
-----------------------------------------------------
--              Subtypes and Constraints
-----------------------------------------------------
```

61. subtype-declaration →
    subtype IDENTIFIER **is** subtype-indication ;

62. subtype-indication →
    [*function*-mark] *type*-mark [constraint]

63. constraint →
    range-constraint
    # index-constraint

64. range-constraint →
    range range-specification

65. index-constraint →
    (discrete-range { , discrete-range })

66. discrete-range →
    subtype-indication
    # range-specification

67. range-specification →
    type-mark ' **range** [(*static*-expression)]
    # expression direction expression

68. direction →
    to
    # **downto**

```
-----------------------------------------------------
--     Objects, Aliases, Files, Disconnections
-----------------------------------------------------
```

69. object-declaration →
    object-class identifier-list : subtype-indication [signal-kind]
    [:= expression] ;

70. object-class →

signal
constant
variable

71. signal-kind →
    bus
    # **register**

72. alias-declaration →
    alias IDENTIFIER : subtype-indication **is** name ;

73. file-declaration →
    file IDENTIFIER : subtype-indication **is** [mode]
    string-expression ;

74. disconnection-specification →
    disconnect signal-list : *type*-mark **after**
    time-expression ;

75. signal-list →
    name { , name }
    # **others**
    # **all**

--   Attribute Declarations and Specifications
----

76. attribute-declaration →
    attribute IDENTIFIER : *type*-mark ;

77. attribute-specification →
    attribute IDENTIFIER **of** entity-specification **is**
    expression ;

78. entity-specification →
    entity-name-list : entity-class

79. entity-name-list →
    designator { , designator }
    # **others**
    # **all**

80. entity-class →
    entity
    # **architecture**
    # **package**

```
                    # configuration
                    # component
                    # label
                    # type
                    # subtype
                    # procedure
                    # function
                    # signal
                    # variable
                    # constant
```

```
           -----------------------------------------------
           --              Schemes
           -----------------------------------------------
```

81. generation-scheme →
        if-scheme
        # for-scheme

82. iteration-scheme →
        for-scheme
        # while-scheme

83. if-scheme →
        if *boolean*-expression

84. for-scheme →
        for IDENTIFIER **in** discrete-range

85. while-scheme →
        while *boolean*-expression

```
           -----------------------------------------------
           --              Concurrent Statements
           -----------------------------------------------
```

86. concurrent-statements →
        { concurrent-statement }

87. concurrent-statement →
        block-statement
        # concurrent-assertion-statement
        # concurrent-procedure-call
        # concurrent-signal-assignment-statement
        # component-instantiation-statement
        # generate-statement
        # process-statement

88. block-statement →
    IDENTIFIER :
    block [(*boolean*-expression)]
      [**generic** interface-list ;]
      [**generic map** association-list ;]
      [**port** interface-list ;]
      [**port map** association-list ;]
    begin
      concurrent-statements
    end block [IDENTIFIER] ;

89. component-instantiation-statement →
    IDENTIFIER : *component*-mark
      [**generic map** association-list]
      [**port map** association-list] ;

90. concurrent-assertion-statement →
    [IDENTIFIER :] assertion-statement

91. concurrent-procedure-call →
    [IDENTIFIER :] procedure-call-statement

92. concurrent-signal-assignment-statement →
    [IDENTIFIER :] conditional-signal-assignment
    # [IDENTIFIER :] selected-signal-assignment

93. conditional-signal-assignment →
    target <= options conditional-waveforms ;

94. conditional-waveforms →
    { waveform **when** *boolean*-expression **else** }
    waveform

95. waveform →
    waveform-element { , waveform-element }

96. waveform-element →
    expression [**after** *time*-expression]
    # **null** [**after** *time*-expression]

97. target →
    name
    # aggregate

98. options →
    [**guarded**] [**transport**]

99. selected-signal-assignment →
     with expression **select**
         target <= options selected-waveforms ;

100. selected-waveforms →
      { waveform **when** choices , }
      waveform **when** choices

101. generate-statement →
      IDENTIFIER : generation-scheme **generate**
         concurrent-statements
      end generate [IDENTIFIER] ;

102. process-statement →
      [IDENTIFIER :]
      process [(sensitivity-list)]
         { declarative-item }
      begin
         sequential-statements
      end process [IDENTIFIER] ;

103. sensitivity-list →
      name { , name }

```
-------------------------------------------------------
--           Sequential Statements
-------------------------------------------------------
```

104. sequential-statements →
      { sequential-statement }

105. sequential-statement →
      assertion-statement
      # case-statement
      # exit-statement
      # if-statement
      # loop-statement
      # next-statement
      # null-statement
      # procedure-call-statement
      # return-statement
      # signal-assignment-statement
      # variable-assignment-statement
      # wait-statement

106. assertion-statement →

     assert *boolean*-expression
     [**report** *string*-expression]
     [**severity** expression] ;

107. case-statement →
     case expression **is**
       case-statement-alternative
       { case-statement-alternative }
     end case ;

108. case-statement-alternative →
     when choices =>
       sequential-statements

109. exit-statement →
     exit [*label*-IDENTIFIER]
     [**when** *boolean*-expression] ;

110. if-statement →
     if *boolean*-expression **then**
       sequential-statements
     { **elsif** *boolean*-expression **then**
       sequential-statements }
     [**else**
       sequential-statements]
     end if ;

111. loop-statement →
     [IDENTIFIER :]
     [iteration-scheme] **loop**
       sequential-statements
     end loop [IDENTIFIER] ;

112. next-statement →
     next [*label*-IDENTIFIER]
     [**when** *boolean*-expression] ;

113. null-statement →
     null ;

114. procedure-call-statement →
     procedure-mark [association-list] ;

115. return-statement →
     return [expression] ;

116. signal-assignment-statement →

target <= [**transport**] waveform ;

117. variable-assignment-statement →
     target := expression ;

118. wait-statement →
     wait [**on** sensitivity-list]
     [**until** *boolean*-expression]
     [**for** *time*-expression] ;

```
---------------------------------------------------
--        Components and Configurations
---------------------------------------------------
```

119. component-declaration →
     component IDENTIFIER
     [**generic** interface-list ;]
     [**port** interface-list ;]
     end component ;

120. block-configuration →
     for block-specification
     { use-clause }
     { configuration-item }
     end for ;

121. block-specification →
     architecture-mark
     # *block-label*-IDENTIFIER
     # *generate-label*-IDENTIFIER [(index-specification)]

122. index-specification →
     discrete-range
     # *static*-expression

123. configuration-item →
     block-configuration
     # component-configuration

124. component-configuration →
     for component-specification
     [**use** binding-indication ;]
     [block-configuration]
     end for ;

125. configuration-specification →
     for component-specification **use** binding-indication ;

126. component-specification →
    instantiation-list : *component*-mark

127. instantiation-list →
    identifier-list
    # **all**
    # **others**

128. binding-indication →
    entity-aspect
    [**generic map** association-list]
    [**port map** association-list]

129. entity-aspect →
    entity *entity*-mark [(*architecture*-IDENTIFIER)]
    # **configuration** *configuration*-mark
    # **open**

Alphabetical Index of Rules:

Appendix C
Suggested Reading

Armstrong, James R. **Chip-Level Modeling with VHDL.** (Englewood
Cliffs, NJ: Prentice Hall: 1988).

Barton, David *Behavioral Descriptions in VHDL.* VLSI Systems
Design, June 1988.

Barton, David *A First Course in VHDL.* VLSI Systems Design -
Design Automation Guide, January 1988.

Coelho, David **The VHDL Handbook.** (Boston: Kluwer Academic:
scheduled Summer 1989).

Coelho, David *VHDL : A Call for Standards.* 25th ACM/IEEE Design
Automation Conference Proceedings, 1988.

Hines, John *Where VHDL Fits Within the CAD Environment.* 24th
ACM/IEEE Design Automation Conference Proceedings, 1987.

**IEEE Standard VHDL Language Reference Manual - Std 1076-
1987.** (New York: IEEE: 1988).

Saunders, Larry *The IBM VHDL Design System.* 24th ACM/IEEE
Design Automation Conference Proceedings, 1987.

Index